THE LANGUAGE OF LOGIC

THE LANGUAGE OF LOGIC

A Self-Instruction Text

Second Edition

Morton L. Schagrin

State University of New York
at Fredonia

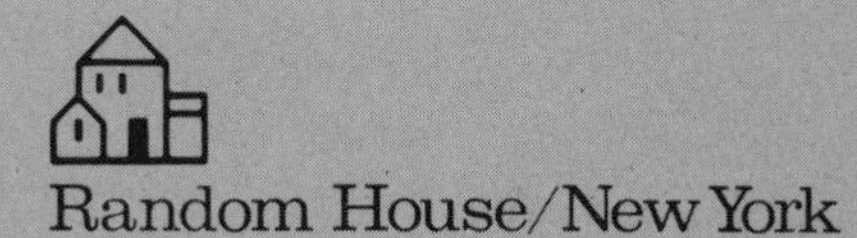

Random House/New York

Second Edition

987654321

 All inquiries should be addressed to Random House, Inc., 201 East 50th Street, New York, N.Y. 10022. Published in the United States by Random House, Inc., and simultaneously in Canada by Random House of Canada Limited, Toronto.

Library of Congress Cataloging in Publication Data
Schagrin, Morton L.
The language of logic.
Includes index.
1. Logic, Symbolic and mathematical—Programed instruction. I. Title.
BC135.S3 1979 511'.3'077 78-9411
ISBN 0-394-31299-6

Cover by Vladimir Yevtikhiev; text design by Deborah Payne

Manufactured in the United States of America

Acknowledgments

The primary debt any programer has is to the long-suffering students who tested the early versions. To them I again extend my gratitude for their assistance. Thanks are also due to the many colleagues, fellow programers, and reviewers whose helpful suggestions have guided me in this work. This work could not have been done without the assistance and cooperation of the Programed Instruction Project of the Great Lakes Colleges Association.

This program was developed in part pursuant to a contract with the United States Office of Education, Department of Health, Education, and Welfare.

Revised Edition

My thanks are due to those who have written me over the past few years with critical comments and helpful suggestions, some of which I have adopted. I am especially grateful to Professor R. B. Angell, who saved me from some horrible mistakes in Part III. Finally, my editor, Jane Cullen, deserves a note of appreciation for encouraging me to undertake this revision, and for her firm guidance and patience.

Preface

The symbolism of modern logic is being used more and more frequently in such diverse fields as mathematics, philosophy, psychology, economics, and others. In order to follow the discussion and arguments that employ this symbolism, students have either taken a formal course in modern logic or struggled through a textbook on the subject. The present book was designed to teach the symbolism of modern logic without becoming too deeply involved in the logical questions and proof techniques that comprise the major part of a course or normal textbook.

Specifically, the major goal of this book is to train the reader to recognize when a symbolic formula symbolizes an English sentence, and conversely when an English sentence interprets a symbolic formula.

This book, then, is designed to give anyone of college ability sufficient familiarity with the symbolism of modern logic so that he or she may understand what is asserted by most authors using this symbolism. In my own course in the philosophy of science, for example, many of the reading assignments and much of the class discussion utilize this symbolism. My students, in general, have not studied modern logic, and so must learn this symbolism. In the past I have had to take class time to present this material. Now this text teaches my students what they need to know, and we can begin right away with the substantive problems of the course.

Although this text was not designed to supplement a course in modern symbolic logic, it would be of assistance to those students in such a course who have difficulty in symbolizing or interpreting. Anyone who uses this book as a supplement to a course in logic would do well to work through Part I only. Later when the course takes up predicate logic (first-order functional logic), he can try his hand at Parts II and III.

The median time for working Part I is about two hours; for Part II, about four hours. Part III has not been tested enough to establish a working time.

Finally, the symbolism dealt with in this book derives from Russell and Whitehead's *Principia Mathematica*. The so-called Polish notation is not used at all, nor are any modal symbols used.

Note to Logicians

Although the use-mention distinction has been observed in this program, no attempt has been made to call the reader's attention to it. To teach this distinction properly, and at the very beginning, would have increased the length of the program unduly. Besides, for many *applications* of the symbolism, the distinction is not critical. This program was not designed to be a rigorous introduction to logical theory. Thus if it is used in conjunction with a course in logic or semantics, some classroom time will have to be devoted to teaching the use-mention distinction.

Instructions to the Reader

You are about to learn some skills. Merely reading about what to do will not make you skillful—try telling someone how to tie a bow knot. Therefore, you should be prepared to do something in order to learn.

Each frame of this program will present some material for you to think about. You will then be asked to perform some task—for example, circle a word, pick an answer, fill in a blank, or match some items. Then either of two things will happen:

(1) The correct answer will be written below the place you performed the task and you will continue on through the frame, or the correct answer will appear at the bottom of the frame and you will be directed to turn to some other frame. OR,

(2) After performing your task, you will be directed to turn to a new frame where you will find the correct answer. Frame numbers are the large numbers at the top of each page.

Whenever you have answered incorrectly, you should examine the correct answer carefully to see why you were wrong. In most cases the discussion following the correct answer will explain why it is the correct one. You may continue the program even if you are not perfectly clear why you were wrong (or right). BUT when the directions sending you appear in a SHADED BOX (for example, Turn to frame 31), you should not continue unless you understand why the correct answer *is* correct. You may even have to go back over the preceding section before you feel you fully understand the answer.

To help you avoid glancing at the answers, ANSWER SHIELDS are provided at various points in the program. These shields should be torn out of the book when you come to them. They can also be used for scratchwork and calculations, and some of them have useful summary notes printed on them. The points in the book where these shields appear are good stopping places when you get tired.

A Self-Examination with which you can test your mastery of the material appears at the end of each part.

Contents

Part 1

Sentence Logic (SL)

Introduction

Over the years natural languages—such languages as English, French, and Russian—have evolved into very rich and versatile social and cultural phenomena. Among their many uses, natural languages provide a setting for arguments, inferences, and reasoning of all kinds. Yet their very richness and flexibility often obscure reasoning processes; the principles justifying various conclusions are often hidden in the labyrinths of the grammar of a natural language. In order to meet the growing demand for clarity, precision, and explicitness in reasoning, logicians have developed a system that makes it easier for us to trace *some* of the connections among those statements that occur in arguments and theories.

This system, called sentence logic (SL), is carefully constructed, with various symbols and patterns of symbols being used to represent well-defined concepts. Sentence logic, then, is an artificial language by means of which one can present arguments and theories. Presented in this form, arguments and theories can be examined and evaluated more rigorously than would be the case if they were stated in some natural language. It is for these reasons that we study sentence logic, although few of us write or think in the symbolism of this artificial language; most of us express our ideas and thoughts by means of the symbols and grammar of our native natural language.

In Part I, you will learn to recognize the symbols used in this artificial language, and by constructing truth tables you will learn what role these symbols play in the language. You will also learn how to identify which symbolic expression in sentence logic correctly represents an English sentence of moderate complexity. You will also learn the reverse of this process: to identify the English sentence that correctly interprets a symbolic formula of sentence logic. Finally, you will learn one important relation that some sentences have to others—namely, logical equivalence. And you will learn how to establish that two sentences of SL are or are not logically equivalent.

To summarize, upon completion of Part I you should be able to

(1) Select from a list of symbols those symbols corresponding to the five most common logical sentence connectives.
(2) Construct truth tables for each of these common sentence connectives.

(3) Select the symbolic formula of SL that correctly symbolizes a given compound English sentence of moderate complexity.
(4) Select an English sentence that correctly interprets a given formula of SL.
(5) Match basic formulas that are logically equivalent.

TEAR THIS SHEET OUT

You may use this sheet as your shield. You may also do scratch work on it.

DO NOT MAKE NOTES.

------------------------- fold here -------------------------

1

English sentences can be separated for logical purposes into those that are *simple* (or atomic) and those that are *compound* (or molecular).

SIMPLE SENTENCES:

a. Small grey doves do coo at lonesome lovers.
b. The corn is quite tall.

COMPOUND SENTENCES:

a. Dogs lope while horses gallop.
b. If the balance of terror persists, then the rate of cigarette smoking increases.

Roughly speaking, compound sentences are composed of two or more shorter simple sentences. CIRCLE the compound sentences in this list:

1. The quick brown fox jumped over the lazy dog.

2. Roses are red and violets are blue.

3. Now is the time for all good men to come to the aid of the party.

4. Fat burns, but water doesn't (burn).

5. He who never speaks never errs.

6. If a man never errs, then the Devil isn't happy.

7. Simple sentences standing alone are easy to detect; nevertheless, it is often difficult to analyze a compound sentence into its component simple parts.

8. Seldom have so many owed so much to so few.

---

2. Roses are red and violets are blue.
4. Fat burns, but water doesn't (burn).
6. If a man never errs, then the Devil isn't happy.
7. Simple sentences standing alone are easy to detect; nevertheless, it is often difficult to analyze a compound sentence into its component simple parts.

If you have missed any of these, turn to frame 3.

If you are correct on all of these, turn to frame 4.

You have not followed directions. You need to cooperate in order for this program to succeed. Follow the directions printed at the bottom of this frame.

Please reread Instructions to the Reader, page ix. Then return to frame 1 and FOLLOW THE DIRECTIONS.

3

You missed one or more compound sentences.

Perhaps you should concentrate on identifying simple sentences. Simple sentences have no shorter sentences contained within them. Take, for example,

Dogs bark.

Modifiers are irrelevant to the logical simplicity of a sentence. Take, for example, the sentence

Small, short-haired black dogs often bark fiercely.

This is still a simple sentence: Some thing (the subject) does something (the predicate).

Were we to compound 'Dogs bark' with another simple sentence, we might obtain something like

If *dogs bark,* then *cats purr.*

Circle the compound sentences below:

1. That doctor is a fraud or this druggist is a charlatan.

2. That child kicked this child in the stomach.

3. Aversive reinforcement is rewarding to a masochist.

4. I will not run if I am nominated, and I shall not serve if I am elected.

---

1. That doctor is a fraud or this druggist is a charlatan.
4. I will not run if I am nominated, and I shall not serve if I am elected.

If you are incorrect, go to frame 12.

If you are correct, go to frame 4.

4

Underline just the simple sentences in the following list, which contains all the compound sentences from frame 1. Do not underline the connecting words.

Example: If ducks bark, then cats quack and dogs purr.

1. Simple sentences standing alone are easy to detect; nevertheless, it is often difficult to analyze a compound sentence into its component simple parts.

2. Roses are red and violets are blue.

3. Fat burns, but water doesn't (burn).

4. If a man never errs, then the Devil isn't happy.

Go to frame 5.

5

You probably have

1. Simple sentences standing alone are easy to detect; nevertheless, it is often difficult to analyze a compound sentence into its component simple parts.
2. Roses are red and violets are blue.
3. Fat burns, but water doesn't.
4. If a man never errs, then the Devil isn't happy.

Let's look more closely at the sentence

The Devil isn't happy.

In a sense, this contains the shorter simple sentence

The Devil is happy.

You may object to the metaphorical term 'contains'. If so, you may prefer to say that 'The Devil isn't happy' is logically related to 'The Devil is happy'. Clearly, if one sentence is true, the other is false, and conversely. When sentences are so related, we shall say that one is the *denial* or *negation* of the other.

Write the simple sentences of which the following are denials:

1. This is not a simple sentence.

   ____

2. Violets aren't blue.

   ____

3. He cannot make it.

   ____

4. It is not the case that the moon is made of green cheese.

   ____

---

1. This is a simple sentence.
2. Violets are blue.
3. He can make it.
4. The moon is made of green cheese.

Go to frame 6.

6

It will save space if, in any problem, we *abbreviate* the simple sentences by using capital letters. For example, '*S*' abbreviates 'This is a simple sentence'. '*B*' abbreviates 'Violets are blue'.

M: He can make it.

G: The moon is made of green cheese.

We can now (partially) symbolize the *negations* of these simple sentences as

not *S*	not *M*
not *B*	it is not the case that *G*

Indeed, we shall go further and write a '~' to the left of any sentence to symbolize its *denial*.

ENGLISH SENTENCE	SYMBOLIC FORMULA
This is not a simple sentence	~*S*
Violets aren't blue.	______
He cannot make it.	______
It is not the case that the moon is made of green cheese.	~*G*

(Fill in the above blank spaces.)

Violets aren't blue: ~*B*
He cannot make it: ~*M*

Go to the next frame.

7

Complete the examples, using '~'.

1. Fat burns, but water does not.
   F: Fat burns.
   W: Water burns. *F* but ~*W* ______

2. It is not the case that both all redheads are communists and all redheads are capitalists.
   C: All redheads are communists.
   A: All redheads are capitalists. ~both *C* and *A* ______

3. Either Jones isn't guilty or I'll turn in my badge.
   J: Jones is guilty.
   B: I'll turn in my badge. Either ______ or ______
   (Fill in.)

---

3. Either ~*J* or *B*

---

4. If dogs bark, then cats don't swim.
   D: Dogs bark.
   C: Cats swim. If *D* then ~*C* ______

5. Although roses are red, violets aren't blue.
   R: Roses are red.
   B: Violets are blue. ______

---

5. Although *R*, ~*B*

---

6. It is not the case that if it rains then the day is ruined.
   R: It rains.
   D: The day is ruined. ______

---

6. ~if *R* then *D*

If you were wrong on *either* 5 or 6, go to frame 8.

If both have been correct, go to frame 9.

8

You are having some trouble placing the negation sign in the right spot. The trick to this is to

(1) find the sentence being denied,
(2) find its abbreviation, or symbolization if it is compound, and
(3) write the symbolized negation in the larger sentence without changing any of the connective words.

Study the following examples:

That *won't* work, *or else* I am a monkey's uncle.
W: That will work.
M: I am a monkey's uncle.

~*W*, or else *M*

It is not the case that either 2 is odd *or* 6 is prime.
O: 2 is odd.
P: 6 is prime.

~either *O* or *P*

(Here the entire sentence 'either *O* or *P*' is denied.)

Let's try some more:

1. Airplanes frighten me *but* sailboats *don't*.
A: Airplanes frighten me.
S: Sailboats frighten me.

____

---

1. *A* but ~*S*

---

2. If you eat ice cream with pickles, then you will not sleep well tonight.
P: You eat ice cream with pickles.
S: You will sleep well tonight.

____

---

2. If *P* then ~*S*

Go to frame 10.

9

Here is one final example:

If sugar is not soluble, then neither is salt.
S: Sugar is soluble.
A: Salt is soluble.

____

If ~*S* then ~*A*

Go to frame 11.

10

Here is one final example:

> If sugar is not soluble, then neither is salt.
> *S*: Sugar is soluble.
> *A*: Salt is soluble.

____

If $\sim S$ then $\sim A$

If you are correct, go to frame 11.

If you are still incorrect, go to frame 12.

11

Different authors use different signs to symbolize *denial*. You should be familiar with some of the common variations. In addition to '∼' , you will find '−' and '¬' .

Of course, any one author will use just one of these. So the *negation* of '*S*' may be written in one of these ways:

$$\sim S \qquad -S \qquad \neg S$$

Write the denial of 'Roses are red' (abbreviated: *R*) in the three different ways indicated above.

____

Do *not* turn to the next frame; go to frame 13.

12

You are here because you are having difficulties with fairly easy but basic material.

Put this book away for a while and do something else, or perhaps have a cup of coffee. Then come back and review the section where you are having trouble. If you still find that you do not understand the material, consult with your instructor or someone who knows some logic.

13

Recall that '~R' (or '¬R' or '−R') is a true sentence if 'R' is false. Denial is the first truth-functional compound we shall learn.

Another compound whose truth-value (true or false) depends on the truth-values of its components is the *conjunction* of two sentences. In English the word 'and' is often used *conjunctively*, although the word 'but' frequently conjoins two sentences.

Circle the word in the last sentence above that is being used to conjoin the two simple sentences there.

Although

Go to frame 14.

14

The English language has many devices to signal conjunction; even punctuation marks can play this role. We shall use '&' between two symbolized sentences to construct their conjunction.

For instance, 'Although roses are red, violets aren't blue', which was abbreviated as 'Although *R*, ~*B*', will henceforth be written '(*R* &~*B*)'.

This symbol for conjunction is not completely standard. You will occasionally find an author who uses '·' or '∧' where we use '&'.

Hence the sentence 'Roses are red and violets are blue' will be variously symbolized as

$$(R \cdot B) \qquad (R \wedge B) \qquad (R \mathbin{\&} B)$$

Circle those formulas that symbolize conjunctions. (Remember that capital letters are abbreviations of complete simple sentences.)

1. $(\sim A \in B)$
2. $(C \mathbin{\&} \sim\sim D)$
3. $(E \square F)$
4. $(G \vee H)$
5. $(J \cdot K)$
6. $(\neg L \wedge M)$
7. $(\sim N + \sim O)$
8. $(-P \mathbin{\&} Q)$
9. $[(R \cdot S) \cdot T]$
10. $(\sim U \mathbin{\&} V)$

---

2. $(C \mathbin{\&} \sim\sim D)$
5. $(J \cdot K)$
6. $(\neg L \wedge M)$
8. $(-P \mathbin{\&} Q)$
9. $[(R \cdot S) \cdot T]$
10. $(\sim U \mathbin{\&} V)$

Go to frame 15.

15

These symbols are never mixed. An author will choose one of them and never use the others.

CORRECT	INCORRECT
$(\sim A \cdot \sim B)$	$(\sim A \cdot -B)$
$[(C \ \& \ D) \ \& \ E]$	$[(C \cdot D) \ \& \ E]$
$[(\neg\neg A \wedge B) \wedge C]$	$[(\neg\sim A \wedge B) \cdot C]$
$[(F \cdot G) \cdot (H \cdot J)]$	$[(F \cdot G) \wedge (H \cdot J)]$

Write the technical names of the two sorts of compound sentences we have learned.

C_______________N

D_______________L or N_______________TION

Go to frame 16.

16

From the list below, select the symbols corresponding to the logical connections, and copy them in the indicated spaces. Try not to check back on this one.

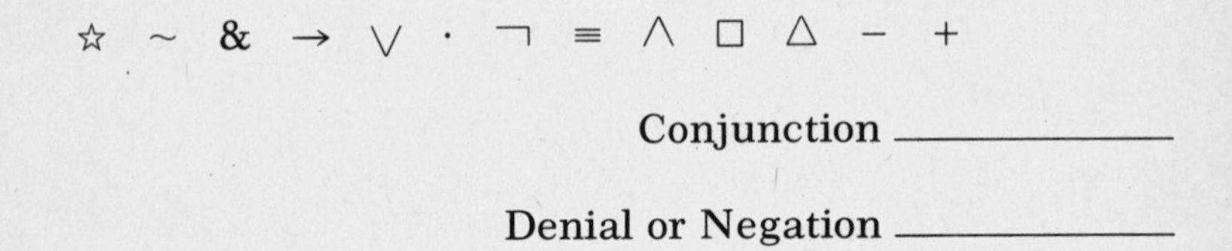

Conjunction ∧ · &
Denial ~ ¬ −

Go to frame 17. But first check p. ix if you don't remember about shaded instructions.

17

We can go on to introduce more symbols, or we can stop now. The latter *alternative* is not as defeatist as it may sound, and you can refer to any modern logic textbook to see how much can be done with only the concepts of denial and conjunction. However, we will go on to take a look at some new symbols.

Underline the connective in the paragraph above for which we have *not* introduced a symbol.

Go to frame 18.

18

EITHER you found it easily OR you are not concentrating on the tasks in this program.

Fortunately there is only one symbol used by everyone for *alternation* or *disjunction*. The first sentence of this page is a disjunction and is symbolized like this:

$(E \vee \sim C)$
E: You found it easily.
C: You are concentrating on the tasks in this program.

Now you try to symbolize a *disjunction*:

This bridge is safe, or Jones is a poor engineer.
B: This bridge is safe.
J: Jones is a poor engineer. ____________________

$(B \vee J)$

Did you forget parentheses? If so, put them in, and everyone go to frame 19.

19

'$(B \vee J)$' asserts that one of two things is the case. Could we maintain the truth of the compound if both alternatives were true? That is, suppose the bridge *is* safe, and still Jones is a poor engineer—is '$(B \vee J)$' true or false?

Lawyers use the term 'and/or' to express the condition that either of two sentences *or possibly both* are true. We shall always use '$\vee$' in the sense of the lawyers' 'and/or'. Hence '$(B \vee J)$' *is a false compound only when 'B' and 'J' are* BOTH *false sentences.*

Now, suppose 'A' is some true sentence (it doesn't matter which), and 'B' is some false sentence. What is the truth-value of '$(A \vee B)$'?

True/False ________

True, go to frame 21.

False, go to frame 20.

20

Perhaps you're not concentrating enough. Read the middle paragraph in frame 19 again. Since at least one *disjunct* is true, the whole disjunction is true.

Return to frame 19.

21

'$(A \vee B)$' is true, since not both disjuncts are false.

Let's try some more. Determine if the indicated compound is true or false, given the truth-values of the component sentences.

1. $(C \vee D)$
   where 'C' is false
   and 'D' is true ______________

2. $(E \vee F)$
   where 'E' is true
   and 'F' is true ______________

3. $(G \vee H)$
   where 'G' is false
   and 'H' is false ______________

4. $(J \vee {\sim}K)$
   where 'J' is false
   and 'K' is false ______________

Only '$(G \vee H)$' is false. In the last example, 'K' is false; therefore its negation '${\sim}K$' is true. Thus, not both disjuncts are false and the whole disjunction is true.

If you missed number 1, 2, or 3, return to frame 19 and study what is said there. Then continue reworking the program.

If you missed number 4, go to frame 22.

If you were completely right, go to frame 23.

22

You did not realize that '$(J \vee \sim K)$' is true when 'J' is false and 'K' is false.

The disjunction, or alternation, is between 'J' and '$\sim K$'. In order to determine the truth-value of the disjunction, we must first determine the truth-value of the disjuncts. The truth-value of 'J' is given to us as false. But the truth-value of '$\sim K$', the other disjunct, must be calculated from the truth-value of 'K'. Since 'K' is false, '$\sim K$' is true. Hence we have a disjunction of a false sentence, 'J', and a true one, '$\sim K$'.

A disjunction is false only when *both* disjuncts are false.

Go to frame 23.

25

'C or R but not both C and R' should be broken up piece by piece. The major break in this sentence occurs at 'but', which in this context signals a conjunction. Hence we have a conjunction of two sentences: 'C or R' and 'not both C and R'.

'C or R' can be symbolized only one way: '$(C \vee R)$'. The second conjunct, 'not both C and R', is the negation of 'both C and R'—not 'both C and R'. Choosing some representative negation and conjunction symbols, we have, for instance, '$\neg(C \wedge R)$'.

Now we must conjoin '$(C \vee R)$' with '$\neg(C \wedge R)$'. Since we have already chosen '$\wedge$' for conjunction, we get '$(C \vee R) \wedge \neg(C \wedge R)$'.

The other correct choices are obtained from the above by changing the symbols throughout for denial and conjunction.

Go to frame 26.

26

The final two connectives we shall study are perhaps the most important from the point of view of deduction techniques. We have already met with examples of *conditional* sentences: they have the form

> If . . . , then ______.

An example of a *biconditional* sentence is

> You will pass the course *if and only if* you pass the final examination.

This means

> If you pass the final examination, then you will pass the course, and if you do not pass the final examination, then you will not pass the course.

There are two accepted ways of symbolizing the *conditional* and the *biconditional.*

CONDITIONAL	BICONDITIONAL
$\supset$	$\equiv$
$\rightarrow$	$\leftrightarrow$

Circle the symbolic formula that symbolizes the given English sentence.

1. *If* it rains, *then* we do *not* go on the picnic *and* we stay home.
   - R: It rains.
   - P: We go on the picnic.
   - S: We stay home.

   $\sim(R \supset P)$ & S

   $R \supset (\sim P \cdot S)$ (circled)

   $R \supset (\neg P \vee S)$

   $(R \supset -P) \cdot S$

2. *If* the wind is *not* too strong *and* it is *not* raining, we can have the race.
   - W: The wind is too strong.
   - R: It is raining.
   - H: We can have the race.

   $(\sim W$ & $\sim R)$ & H

   $\sim W \rightarrow \sim(R \cdot H)$

   $\sim W \supset (\sim R \supset H)$

   $(\sim W \cdot \sim R) \rightarrow H$ (circled)

3. We are staying home, and furthermore, if it is sunny, then we shall miss the swim.

   H: We are staying home.
   S: It is sunny.
   M: We shall miss the swim.

   $(H \supset S) \cdot M$

   $H \supset (S \supset M)$

   $(H \,\&\, S) \rightarrow M$

   $H \cdot (S \rightarrow M)$

---

3. $H \cdot (S \rightarrow M)$

---

4. If Harrison wins the presidency, or if the Socialist party gains a majority, the country is doomed.

   H: Harrison wins the presidency.
   S: The Socialist party gains a majority.
   C: The country is doomed.

   $H \supset (S \vee C)$

   $(H \vee S) \rightarrow C$

   $H \vee \supset (S \supset C$

   $(H \cdot S) \rightarrow C$

---

4. $(H \vee S) \rightarrow C$

   The formula '$H \vee \supset (S \supset C)$' is not well-formed and does not symbolize *any* sentence at all.

If either of these is incorrect, go to frame 27.

If they are both correct, go to frame 28.

27

You are having trouble identifying the 'if' clause and the 'then' clause. These clauses may be compound, or the conditional 'If . . . , then ______' sentence may be compounded with something else.

Try using parentheses on the English sentence.

1. (If you are right then I apologize), but (I think you are wrong.)
   R A W

   $(R \rightarrow A) \cdot W$

2. If c = 7 then (either a = 3 or b = 5).
   C A B

   $C \supset (A \vee B)$

   Circle the correct choice in example 3.

3. If Jones is present and Smith is absent, something important is going on.
   J: Jones is present.
   S: Smith is absent.
   I: Something important is going on.

   $J \supset (S \cdot I)$

   $(J \;\&\; S) \rightarrow I$

   $(J \supset S) \supset I$

   $(J \;\&\; S) \;\&\; I$

---

3. $(J \;\&\; S) \rightarrow I$

If you are correct, return to frame 26 and rework the problems before you continue the program.

If you are incorrect, go to frame 12.

28

Great care must be taken to determine which sentence is the *antecedent*, and which the *consequent* of a conditional compound. For example,

You will find him if you turn right at the corner
H R

is symbolized '$(R \rightarrow H)$'.

Furthermore, the phrase 'only if' is highly deceptive. For example,

You are able to enter only if you have a ticket

may be rephrased which way? Underline one.

1. If you have a ticket, then you are able to enter.

2. If you are able to enter, then you have a ticket.

Choice 1, go to frame 29.
Choice 2, go to frame 30.

29

The sentence 'You are able to enter *only if* you have a ticket' means that your having a ticket is *necessary* for you to enter, but it may not be sufficient. After all, even if you have your ticket, the theater may be filled, you may not be properly dressed, you could be drunk, and so forth. Therefore, 'If you are able to enter, then (at least) you have a ticket' is closer in meaning to our original sentence.

Go on to the next frame.

30

You said 'You are able to enter only if you have a ticket' may be rephrased as 'If you are able to enter, then you have a ticket'. Correct; you realized that even if you have a ticket, you still might not be able to enter (for example, the theater might be filled).

Happily, these problems do not arise with the biconditional ($\equiv$, $\leftrightarrow$). There is no need to distinguish the left-hand sentence from the right-hand one, as we had to distinguish antecedent from consequent in a conditional.

Symbolize the sentence

If and only if the time is ripe, will the revolution succeed.

T: The time is ripe.

R: The revolution will succeed. ______________________

(Answer)

Any of these four are correct:

$(T \equiv R)$ $(T \leftrightarrow R)$

$(R \equiv T)$ $(R \leftrightarrow T)$

Go on to frame 31.

31

The phrase 'if and only if' is used mostly in technical writing. The same idea is expressed more casually by the phrase 'when and only when' as in the sentence

> These rules may be disregarded when and only when collision is imminent.

We might find an instruction that reads

> Use this door in case there is a fire.

This might be symbolized as '$(F \rightarrow D)$'.

Choose the correct symbolization for the instruction

> Use this door *just in case* there is a fire.
>
> 1. $(F \rightarrow D)$
> 2. $(D \rightarrow F)$
> 3. $(F \leftrightarrow D)$

---

3. $(F \leftrightarrow D)$ and of course $(D \leftrightarrow F)$

Go to the next frame.

32

From the list below, select the symbols corresponding to the logical connections and copy them in the indicated spaces.

≡ & ☆ ~ → □ ↔ ∨ – · △ ⊃ ∧ ¬ ° +

Denial ________________

Conditional ________________

Conjunction ________________

Biconditional ________________

Disjunction ________________

Denial ~ ¬ –
Conditional ⊃ →
Conjunction & · ∧
Biconditional ≡ ↔
Disjunction ∨

Go to p. 41.

TEAR THIS SHEET OUT

You may use this sheet as your new shield and as a reference sheet. You may also do scratch work on it. DO NOT MAKE NOTES.

Denial $\sim$ $\neg$ $-$

Conjunction $\wedge$ & $\cdot$

Disjunction $\vee$

Conditional $\rightarrow$ $\supset$

Biconditional $\leftrightarrow$ $\equiv$

------------------------- **fold here** -------------------------

33

We shall now learn how the truth-value of a compound is a *function* of the truth-values of its components. Since we wish to be quite general about this, we shall not use *actual* sentences which are in fact true or false but rather *sentential variables,* such as '*p*', '*q*', '*r*', and '*s*'.

Using these lower-case letters (and others if we need them), we can avoid such tedious and turgid remarks as

> The denial of any sentence is symbolized by a tilde, '~', written immediately to the left of the symbolization of that sentence.

Using a variable, we can say instead "The denial of any sentence p is $\sim p$." *Instances* of this usage are

PARTICULAR SENTENCES	THEIR DENIALS
B	$\sim B$
$(C \vee D)$	$\sim(C \vee D)$
$[(E \supset F) \equiv G]$	$\sim[(E \supset F) \equiv G]$
$\sim H$	$\sim\sim H$

Letting 'T' and 'F' abbreviate 'true' and 'false' respectively, we can show in tabular form how the truth-value of a denial is a function of the truth-value of the sentence denied:

p	$\sim p$
T	F
F	T

This *truth table* shows that for *any* sentence p,
when p is T, $\sim p$ is ______________
when p is F, $\sim p$ is ______________
(Fill in the blanks.)

---

when p is T, $\sim p$ is F
when p is F, $\sim p$ is T

Go to frame 34.

34

We have learned that a disjunction, $(p \vee q)$, is false only when both disjuncts are false, and the compound is true in any other case. For *any* two *arbitrary* sentences p and q, we can list all possible truth-value combinations of them:

	p	q	
Case 1. both true	T	T	
Case 2. p true; q false	T	F	
Case 3. p false; q true	F	T	
Case 4. both false	F	F	

Only in case 4 is the disjunction of these two sentences false. We write 'F' for that case:

p	q	$(p \vee q)$
T	T	
T	F	
F	T	
F	F	F

Now construct the complete truth table for $(p \vee q)$.

p	q	$(p \vee q)$
T	T	T
T	F	
F	T	
F	F	

Go to the next frame.

35

A disjunction is false only when both disjuncts are false; in all other cases, the disjunction is true.

p	q	$(p \vee q)$
T	T	T
T	F	T
F	T	T
F	F	F

What about conjunctions? When I assert p and q conjointly, I intend that both p and q are the case. If either p, or q, or both p and q, are false, my conjoint assertion of them, (p and q), is likewise false.

Complete the truth table for conjunction.

	p	q	$(p \& q)$
Row 1	T		
Row 2	T		F
Row 3	F	T	
Row 4	F	F	

Go to the next frame.

36

Compare the truth table for *conjunction* with that for *disjunction;* note the symmetry.

p	q	$(p \cdot q)$	$(p \vee q)$
T	T	T	T
T	F	F	T
F	T	F	T
F	F	F	F

This symmetry is important both for remembering the significance of the concepts of conjunction and disjunction, and for analyzing more complicated compound sentences.

Remember, p and q were understood to be *any arbitrary* sentences, either simple or compound. We can analyze complicated compound sentences by means of truth tables and tell under what conditions these complicated sentences *would be* true or false.

p	$\sim p$	$\sim\sim p$	$\sim\sim\sim p$	$\sim\sim\sim\sim p$
T	F			T
F	T			F

From this table we see that $\sim\sim\sim\sim p$ is F when p is F. Complete the table and determine when $\sim\sim p$ is F and when $\sim\sim\sim p$ is T.

If you have written

Turn to

1. $\sim\sim p$ is F when p is T, and $\sim\sim\sim p$ is T when p is F. frame 37.

2. $\sim\sim p$ is F when p is F, and $\sim\sim\sim p$ is T when p is T. frame 38.

3. $\sim\sim p$ is F when p is F, and $\sim\sim\sim p$ is T when p is F. frame 39.

4. $\sim\sim p$ is F when p is T, and $\sim\sim\sim p$ is T when p is T. frame 40.

37

You have made a mistake. You can go back and try again, or go to frame 38 for some explanation of *why* you have erred.

38

You've made a mistake somewhere, or perhaps you are confused about the *use* of truth tables. A little common-sense reasoning should have warned you about your error. After all, $\sim\sim\sim p$ is the denial of $\sim\sim p$. Therefore, $\sim\sim\sim p$ is true and $\sim\sim p$ is false for some *one* truth-value of p, either T or F. So the correct answer is either 3 or 4.

Go back to frame 36 and work out the truth table again. Then select answer 3 or 4.

But if you want more help, go to frame 40.

39

Choice 3 is correct. From the truth table we see that $\sim\sim p$ is F in the bottom row, which is when p (the far left column) is F. Similarly for $\sim\sim\sim p$.

In order to find the conditions under which a given compound sentence would be true or false, we first must determine the *logical form* of the sentence. This is easily done by *replacing the actual component simple sentences by variables.*

Complete the following examples. Note, however, that the actual sentences abbreviated by the capital letters are not important.

1. $\sim\sim(A \cdot \sim B)$ $\sim\sim(p \cdot \sim q)$ ______
2. $E \supset [M \vee \neg(E \ \& \ G)]$ $p \supset [q \vee \neg(p \ \& \ r)]$ ______

   (Notice that *each* occurrence of 'E' is replaced by 'p'.)
3. $K \wedge (L \wedge S)$ $p \wedge (q \wedge r)$ ______
4. $H \leftrightarrow (J \rightarrow L)$ ______
5. $\sim(\sim A \vee \sim B)$ ______
6. $C \ \& -D$ ______
7. $K \leftrightarrow (K \leftrightarrow K)$ ______

---

4. $p \leftrightarrow (q \rightarrow r)$
5. $\sim(\sim p \vee \sim q)$
6. $p \ \& -q$
7. $p \leftrightarrow (p \leftrightarrow p)$

If you have any wrong, you did not understand the directions. Leaving the parentheses and connective symbols untouched, simply replace each capital letter by a variable. If a capital letter occurs more than once, use the same variable at each occurrence. Go on to frame 41.

40

Something's wrong. Check this truth table against yours in frame 36.

p	$\sim p$	$\sim\sim p$	$\sim\sim\sim p$	$\sim\sim\sim\sim p$
T	F	T	F	T
F	T	F	T	F

The top row alternates T and F; the bottom row alternates F and T.

Looking at the top of the columns, we see that each formula is followed on the *right* by its denial. Thus the truth-value is reversed at each step to the right.

We see then that $\sim\sim p$ is F (bottom row) when, reading to the left, $\sim p$ is T and p is F. So when *any arbitrary* sentence p is F, its *double negation,* $\sim\sim p$, is also F.

Similarly, $\sim\sim\sim p$ is T (still bottom row) when $\sim\sim p$ is F; hence a triple negation is T when the original sentence is F.

Study the construction of the truth table at the top of this page and then turn back to frame 39.

41

We can now determine by truth tables when any compound sentence of a given logical form is true or when it is false. *We simply construct the sentence from its progressively more complex parts.*

Take, for example, the sentence

We will not go to the movies and we will not go to the party.

This is symbolized by the expression '$(\sim M \cdot \sim G)$'. Its logical form is: $(\sim p \cdot \sim q)$.

We observe that we are *conjoining* $\sim p$ and $\sim q$—hence

	p	q	$\sim p$	$\sim q$	$(\sim p \cdot \sim q)$
Row 1	T	T	F	F	
Row 2	T	F	F	T	F
Row 3	F	T	T	F	F
Row 4	F	F	T	T	

Since a conjunction is true only when both conjuncts are true, you should be able to complete the above table. Do so.

---

p	q	$(\sim p \cdot \sim q)$
T	T	F
T	F	F
F	T	F
F	F	T

If you are not sure what you are doing, or have made a mistake in the above table, turn to frame 42. Otherwise go to frame 43.

42

We are working with '$\sim M \cdot \sim G$'. We want to know the conditions under which such a sentence is true (or false). But this sentence is clearly the conjunction of two shorter sentences, and we know the truth conditions for a conjunction: *A conjunction is true only when both conjuncts are true.*

Now our problem is to determine under what conditions both conjuncts are simultaneously true. The conjuncts here are '$\sim M$' and '$\sim G$'. Each of these is true when the simple sentences 'M' and 'G' are false. And *only then.*

All this is summarized neatly in our truth table, where for two arbitrary *simple* sentences, we first determine the truth-values of their negations ($\sim p$ and $\sim q$) and then, following the arrows, the truth-values (for all possible cases) of the conjunction of these negations, ($\sim p \cdot \sim q$).

We'll get more practice with this in a moment. Meanwhile, turn to frame 43.

43

Fill in the following truth tables:

p	$\sim p$
T	

p	q	$(p \& q)$

p	q	$(p \vee q)$

The answers are in the next frame.

44

1. If a sentence is true, its *negation* is false, and if a sentence is false, its negation is true.

p	$\sim p$
T	F
F	T

2. A *conjunction* is true if both components (conjuncts) are true; otherwise it is false.

p	q	$(p \& q)$
T	T	T
T	F	F
F	T	F
F	F	F

3. A *disjunction* or *alternation* is false if both components (disjuncts) are false; otherwise it is true.

p	q	$(p \vee q)$
T	T	T
T	F	T
F	T	T
F	F	F

Go to the next page.

TEAR THIS SHEET OUT

You may use this sheet as your new shield and as a reference sheet. You may also do scratch work on it. DO NOT MAKE NOTES.

Denial $\sim$ $\neg$ $-$

Conjunction & $\wedge$ $\cdot$

Disjunction $\vee$

Conditional $\supset$ $\rightarrow$

Biconditional $\equiv$ $\leftrightarrow$

p	$\sim p$
T	F
F	T

p	q	$(p \,\&\, q)$	$(p \vee q)$
T	T	T	T
T	F	F	T
F	T	F	T
F	F	F	F

------------------------ fold here ------------------------

45

When we assert a conditional such as

> If the temperature exceeds 500° C, the sample will melt

under what conditions are we willing to admit that our assertion is false? Under what conditions may we claim truth for it?

Actually, for some conditionals, there is room for discussion as to what precisely is meant. Just as we took the "weakest" reading for 'or', so we will now take the weakest sense of an 'If . . . , then ______' sentence.

At *the very least*, when we say

(1) If the temperature exceeds 500° C, the sample will melt

we mean to *deny* that

(2) The temperature exceeds 500° C and the sample does not melt.

Let

T: The temperature exceeds 500° C.
S: The sample melts.

Now on the one hand we are asserting (1),(If T then S): $(T \rightarrow S)$, and on the other hand, we are *denying* (2), (T and not S): $(T \cdot \neg S)$.

We shall let the truth table for any conditional $(p \rightarrow q)$ be defined by the truth table for the *denial* of $(p \cdot \neg q)$.

	p	q	$\neg q$	$p \cdot \neg q$	$\neg(p \cdot \neg q)$
Row 1	T	T	F	F	T
Row 2	T	F	T	T	F
Row 3	F	T	F	F	T
Row 4	F	F	T	F	T

By our decision, $(p \rightarrow q)$ is false in one and only one case. For what truth-values of p and q is $(p \rightarrow q)$ false?

Write in 'T' or 'F' as appropriate: p______; q______

Go to the next frame.

46

The conditional $(p \rightarrow q)$ is false only when p is T and q is F.

We have already expressed the intent of a *biconditional* without using the phrase 'if and only if' (sometimes written 'iff'). Recall that the sentence

> You will pass the course if and only if you pass the final examination

whose logical form is $(q \equiv p)$ was expressed by another sentence

> If you pass the final examination, then you will pass the course, and if you do not pass the final examination, then you will not pass the course

whose logical form is: $(p \supset q)$ & $(\sim p \supset \sim q)$.

p	q	$p \supset q$	$\sim p$	$\sim q$	$\sim p \supset \sim q$	$(p \supset q)$ & $(\sim p \supset \sim q)$
T	T	T		F	T	
T	F	F		T	T	
F	T	T		F	F	
F	F	T		T	T	

conjunction of these two

Fill in the truth table. What are the truth conditions for a biconditional?

$(p \leftrightarrow q)$ is true when ______________________________

____

$(p \leftrightarrow q)$ is false when ______________________________

____

Go to the next frame.

47

A biconditional is true if both the right and left components have the same truth-value (i.e., true or false), and a biconditional is false if its major components differ in truth-value.

p	q	$p \supset q$	$\sim p$	$\sim q$	$\sim p \supset \sim q$	$(p \supset q)$ & $(\sim p \supset \sim q)$	$p \equiv q$
T	T	T	F	F	T	T	T
T	F	F	F	T	T	F	F
F	T	T	T	F	F	F	F
F	F	T	T	T	T	T	T

Check this table with the one you filled out in the previous frame.

If there are any errors, go to frame 48.

If you are entirely correct, go to frame 49.

48

p	q	$p \rightarrow q$	$\sim p$	$\sim q$	$\sim p \rightarrow \sim q$	$(p \rightarrow q)\ \&\ (\sim p \rightarrow \sim q)$
T	T					
T	F					
F	T					
F	F					

1. The column under '$\sim p$' exhibits the truth-value of the negation of p. By reference to the column under 'p', we can see how the truth-value of $\sim p$ is a function of the truth-value of p. Similarly for $\sim q$. Fill in these two columns.

2. The column under '$(\sim p \rightarrow \sim q)$' exhibits the truth-value of this *conditional*, and shows how the truth-value is a function of the truth-values of p and q. In constructing this column, we note when the antecedent, $\sim p$, is T and the consequent, $\sim q$, is F. That is the only time $(\sim p \rightarrow \sim q)$ is F. This occurs in *row* 3. Fill in the column.

3. The column under '$(p \rightarrow q)\ \&\ (\sim p \rightarrow \sim q)$' exhibits the truth-value of this *conjunction* as a function of the truth-values of p and q. In constructing this column, we use the *concept of conjunction* whereby a conjunction is T only when both conjuncts [in this case, $(p \rightarrow q)$ and $(\sim p \rightarrow \sim q)$] are T. This occurs in rows 1 and 4. Fill in the last column.

Go to the next frame.

49

Another way to rephrase a biconditional is preferred by some people. The biconditional

p if and only if q

is often reexpressed as

if p then q, and if q then p

symbolized: $(p \to q) \& (q \to p)$.

To show that $(p \leftrightarrow q)$ is *logically equivalent* to $(p \to q) \& (q \to p)$, we need show only that *their truth tables are identical.* That is, we show two sentences to be logically equivalent if we show that under the same conditions both are true or both are false.

p	q	$p \to q$	$q \to p$	$(p \to q) \& (q \to p)$	$p \leftrightarrow q$
T	T				T
T	F		T		F
F	T				F
F	F				T

Fill in the above truth table, taking care to distinguish the antecedent and consequent in $(q \to p)$.

---

p	q	$p \to q$	$q \to p$	$(p \to q) \& (q \to p)$	$p \leftrightarrow q$
T	T	T	T	T	T
T	F	F	T	F	F
F	T	T	F	F	F
F	F	T	T	T	T

If you are correct, go to frame 51.

If you made any mistakes, correct them, and then go to frame 51.

50

Are you getting tired? You missed the last instruction.

Please do not skim through this program. You will learn from it only if you do the tasks carefully and conscientiously and read the directions when you have completed the tasks.

51

You should now be able to construct a truth table for the conditional and the biconditional without any further help.

p	q	$p \rightarrow q$	$p \leftrightarrow q$

---

p	q	$p \rightarrow q$	$p \leftrightarrow q$
T	T	T	T
T	F	F	F
F	T	T	F
F	F	T	T

If you are correct, go to frame 53.

If there are any errors, or if you are unsure of the concepts of the conditional and biconditional, turn to the next frame.

52

The essential point of a conditional compound sentence is that it is *false only when the antecedent is true and the consequent is false.* In a truth table for the conditional $(p \rightarrow q)$, the role of the antecedent is played by p; for $(q \rightarrow p)$, the antecedent is q; and for $(p \vee q) \rightarrow (r \ \& \ s)$, the antecedent is $(p \vee q)$ and the consequent is $(r \ \& \ s)$.

So if we were constructing a truth table for $(p \vee q) \rightarrow (r \ \& \ s)$, we would enter F in any row where $(p \vee q)$ was T and $(r \ \& \ s)$ was F. And if we were constructing a table for $(\sim p \rightarrow \sim q)$, we would enter F in any row where $\sim p$ was T and $\sim q$ was F.

A biconditional is *true when both components have the same truth-value.* Thus $(p \vee q) \leftrightarrow (r \ \& \ s)$ would have T placed in any row where either $(p \vee q)$ and $(r \ \& \ s)$ were T, or $(p \vee q)$ and $(r \ \& \ s)$ were F. For simply $(p \leftrightarrow q)$, those rows where both p and q are T, *or* both are F, have a T in them.

	p	q	r	$(q \vee r)$	$p \rightarrow (q \vee r)$	$p \ \& \ q$	$\sim q$	$\sim q \ \& \ r$	$(p \ \& \ q) \leftrightarrow (\sim q \ \& \ r)$
1.	T	T	T						
2.	T	T	F						
3.	T	F	T						
4.	T	F	F						
5.	F	T	T						
6.	F	T	F						
7.	F	F	T						
8.	F	F	F						

a. For what row(s) is $p \rightarrow (q \vee r)$ false? __________

b. For what row(s) is $(p \ \& \ q) \leftrightarrow (\sim q \ \& \ r)$ true? __________

---

a. $p \rightarrow (q \vee r)$ is false only in row 4.
b. $(p \ \& \ q) \leftrightarrow (\sim q \ \& \ r)$ is true in rows 4, 5, 6, and 8.

Go to the next frame.

53

Although variables, in general, may "stand for" any sentence of any complexity, we have been using variables to replace *simple* sentences within a compound in order to discover the logical form of the compound sentence.

How do you feel about this procedure?

1. I think I know how to find the *logical form* of a sentence, and if that's all I need to know to accomplish the objectives of this program, I'd like to continue. But could I have a little more practice finding the logical forms of sentences?

2. I see how to find the *logical form* of a sentence, but it seems like a lot of extra work. Why can't I just make a truth table with the actual sentences? Why do I have to make replacements with variables?

3. I'm sorry, but I'm still confused about variables and the notion of *logical form*.

Choice 1, turn to page 67.

Choice 2, turn to frame 54.

Choice 3, review material in frames 33–35, and then turn to page 67.

54

Variables enable us to deal with the modality of *possibility*. Consider some actual compound sentence containing the simple sentence

> *S*: The temperature of the sun is extremely high.

If we were to consider the truth conditions of the compound sentence without using variables, we should be forced to say something like

> If '*S*' *were* false, then . . .

Now the subjunctive mood or the concept of contrary-to-fact suppositions is in greater need of clarification than matter-of-fact, declarative assertions, for these assertions are clearly not simply truth-functional. Indeed, we hope that the analysis we are learning in this program will be of assistance in clarifying the logic of intensional and modal discourse.

Thus variables allow us to avoid such locutions as "If '*S*' *were* false, . . ." and to say, rather, "If *q is* false, . . ."

A truth table, then, is *not* a catalog of possibilities in some non-actual world, but is instead a catalog of the results of substituting *actual* true or false sentences for variables.

We shall continue, however, to *speak* informally and to use the subjunctive mood where appropriate.

Turn to page 67.

TEAR THIS SHEET OUT

You may use this sheet as your shield. You may also do scratch work on it.
DO NOT MAKE NOTES.

------------------------- fold here -------------------------

55

Some review is in order before we go on. Fill in the following truth table.

p	q	$(p \vee q)$	$(p \,\&\, q)$

---

p	q	$(p \vee q)$	$(p \,\&\, q)$
T	T	T	T
T	F	T	F
F	T	T	F
F	F	F	F

If you missed any of these, return to frames 34–36 for review and then go to frame 56.

If you are correct, turn to the next frame.

56

Fill in the truth table:

p	q	$p \rightarrow q$	$p \leftrightarrow q$

---

p	q	$p \rightarrow q$	$p \leftrightarrow q$
T	T	T	T
T	F	F	F
F	T	T	F
F	F	T	T

If you missed any of these, return to frames 45–49 for review, and then go on to frame 57.

If you are correct, turn to the next frame.

57

SUMMARY

You have learned about truth-functional sentences, whose truth-values (either true or false) are determined by the truth-values of their component sentences.

There are five common truth-functional connectives, several various symbolizations of these, and several various English words that serve to connect sentences into truth-functional compound sentences.

The significance of truth-functional connections is completely given by the truth-table array of truth-values.

Finally, parentheses are used to indicate how sentences are to be grouped together. Clearly

$\sim(p \rightarrow q)$ is different from $(\sim p \rightarrow q)$.

p	q	$(p \rightarrow q)$	$\sim p$	$\sim(p \rightarrow q)$	$(\sim p \rightarrow q)$
T	T	T	F	F	
T	F	F	F		T
F	T	T	T		
F	F	T	T		

Fill in the blanks. Notice that

$\sim(p \rightarrow q)$ is the denial of a conditional, and

$(\sim p \rightarrow q)$ is a conditional whose antecedent is a denial.

Go to the next frame.

58

Any compound sentence whose truth-value depends only on the truth-values of its components can be symbolized by a careful combination of these five connectives. Which connectives to use, however, is not always indicated by the presence of 'and', 'or', and 'if . . . , then ______'. We must ask ourselves under what conditions the compound would be true or false, and then select our symbols accordingly.

Example, the sentence

> We shall leave at eight o'clock, *unless* it rains.
> L: We shall leave at eight o'clock.
> R: It rains.

Here the truth conditions are not clear. Some people would read this, *weakly*, as

> *If* it does *not* rain, *then* we shall leave at eight o'clock.

This is symbolized by '$\sim R \rightarrow L$'.

Fill in the truth table:

p	q	$\sim p$	$\sim p \rightarrow q$
T	T		
T	F		
F	T		
F	F		

---

p	q	$\sim p$	$\sim p \rightarrow q$
T	T	F	T
T	F	F	T
F	T	T	T
F	F	T	F

If you have made any errors, turn to frame 59.

If this is correct, turn to frame 60.

59

Perhaps it's time for a coffee break.

You should have had no trouble in seeing that in $(\sim p \rightarrow q)$, the antecedent is $\sim p$ and the consequent is q. The antecedent is T and the consequent F *only* in row 4, and so $(\sim p \rightarrow q)$ is F *only* in row 4.

You might refer back to frame 52 before going on to frame 60.

60

We have rephrased 'We shall leave at eight o'clock, unless it rains' as 'If it does not rain, then we shall leave at eight o'clock'.

The truth table for sentences whose form is that of '$\sim R \rightarrow L$' is

	p	q	$\sim p \rightarrow q$
Row 1	T	T	T
Row 2	T	F	T
Row 3	F	T	T
Row 4	F	F	F

Now consider those possible worlds where it is true that *it rains* (and false that it does not rain). What can we infer about our leaving at eight o'clock? That is, assuming that our compound sentence is true and also that it rains, shall we or shall we not leave at eight o'clock?

Which row (or rows) of the truth table is relevant to this question?

____

If you have row 4, turn to frame 61.
rows 1 and 2, turn to frame 62.
rows 3 and 4, turn to frame 63.
rows 1, 2, and 3, turn to frame 64.

61

You said row 4 is relevant. Yet in row 4, p is false, contrary to our assumption that the sentence 'It rains' is true.

Notice also that the entire compound sentence is false in row 4, contrary to our other assumption.

The correct answer cannot include row 4.

Return to frame 60 and choose another answer.

62

Rows 1 and 2 are the relevant ones. CORRECT.

The compound sentence is T in these cases and p ('It rains') is also T in these cases.

We note that q is T in row 1 and F in row 2. This means 'We shall leave at eight o'clock' may be either T or F, and hence it is *undetermined* whether or not we leave at eight o'clock. We may leave at eight o'clock *in spite of* the rain. All we have committed ourselves to in saying, "If it does not rain, then we shall leave at eight o'clock," is *if it does not rain, then we shall surely leave at eight o'clock.*

Example continued from frame 58:

Others would take the assertion of 'L unless R' in the much stronger sense of 'We shall leave at eight o'clock *if and only if* it does *not* rain'. This is symbolized by '$(L \sim R)$'.

Here, if it does rain, then we shall *not* leave at eight o'clock; the decision has been made.

Which row of the following truth table shows us that if it rains we shall *not* leave at eight o'clock?

____

	L	R	$(L \leftrightarrow \sim R)$
	p	q	$p \leftrightarrow \sim q$
Row 1	T	T	F
Row 2	T	F	T
Row 3	F	T	T
Row 4	F	F	F

Turn to frame 65.

63

You said rows 3 and 4. In row 4, the *compound* sentence is F, contrary to our assumption!

Also p is a false sentence in these rows. Thus 'It rains' is a false sentence in these cases. The question, however, assumed that it was true that it rained.

Return to frame 60 and try a different answer, after you have reviewed the page carefully.

64

You said rows 1, 2, and 3. Now, the compound sentence is true in all these cases. *But* in row 3, the truth-value of p is F. We are assuming that 'It rains' is true!

Return to frame 60 and figure out a new answer.

65

Row 3 represents the case where 'L' is false and 'R' is true. Hence, this is the situation where it *does* rain and we do *not* leave at eight o'clock.

We have rephrased 'L unless R' in two different ways:

$(\sim R \rightarrow L)$ $\qquad\qquad$ $(L \leftrightarrow \sim R)$

Which is the *correct* symbolization of a sentence containing the word 'unless'? Certainly no blanket answer can be given. The word 'unless' is used in both a weak and a strong sense. A similar situation arose with the word 'or'. Only the context or further inquiry can reveal the correct symbolization.

Inspect the truth table below and then give a reason why 'L unless R' should not be symbolized as '$R \rightarrow \sim L$'.

$R \rightarrow \sim L$

p	q	$q \rightarrow \sim p$
T	T	F
T	F	T
F	T	T
F	F	T

HINT: Pay special attention to rows 2 and 4.

____

____

____

If 'R' is false (that is, it does not rain), then when the compound sentence is true, we cannot determine if 'L' is true or false. Surely we intend 'L' to be true if it does *not* rain.

Go to the next frame.

66

Use the same symbolization:

L: We leave at eight o'clock.
R: It rains.

Circle the correct symbolization of

We shall leave at eight o'clock whether or not it rains.

1. $L \vee \sim R$
2. $L \rightarrow -R$
3. $(R \vee \sim R) \vee L$
4. $(R \vee \sim R) \rightarrow L$

---

4. $(R \vee \sim R) \rightarrow L$

Turn to the next frame.

67

Circle the formula below that correctly symbolizes the English sentence

> Water boils when and only when the temperature is above 100° C and the vapor pressure is less than 76 cm of mercury.
>
> W: Water boils.
> T: The temperature is above 100° C.
> V: The vapor pressure is less than 76 cm of mercury.

$(T \rightarrow W) \& V$

$W \rightarrow (T \& V)$

$(W \leftrightarrow T) \& V$

$W \leftrightarrow (T \& V)$

$W \leftrightarrow (T \& V)$

If you are incorrect, turn to frame 68.

If you are correct, turn to frame 69.

68

Let's put parentheses around the simple components of the compound sentence and underline connective words.

> (Water boils) <u>when and only when</u> (the temperature is above 100° C) <u>and</u> (the vapor pressure is less than 76 cm of mercury.)

Now replace the English clauses with their abbreviations:

> *W* <u>when and only when</u> *T* <u>and</u> *V*.

The sentence asserts that *W when and only when* two conditions are *jointly* met. Thus the major connective is 'when and only when'. Reflection on the truth conditions of this assertion leads to '$W \leftrightarrow (T \,\&\, V)$'.

Turn to the next frame.

69

Circle the correct formula below for the English sentence

Cheating is not morally right, and if Kant is to be believed neither is lying.

C: Cheating is morally right.
K: Kant is to be believed.
L: Lying is morally right.

$\sim[C \& (K \rightarrow L)]$

$(\sim C \& K) \vee L$

$\sim C \& (K \rightarrow \sim L)$

$(K \rightarrow \sim C) \& \sim L$

$\sim C \& (K \rightarrow \sim L)$

If you are incorrect, turn to the next frame.

If you are correct, turn to frame 71.

70

Perhaps we should rephrase our given sentence

> Cheating is not morally right, and if Kant is to be believed neither is lying

in a more explicit form:

> Cheating is not morally right, and if Kant is to be believed *then lying is not morally right.*

Now place parentheses around the English sentences that are components of this compound:

> (Cheating is not morally right), and if (Kant is to be believed) then (lying is not morally right).

Substituting abbreviations (and negation signs), we get

> $\sim C$, and if K then $\sim L$

This is a conjunction of two statements, '$\sim C$' and 'If K then $\sim L$'.

> $\sim C \ \& \ (K \rightarrow \sim L)$

Turn to the next frame.

71

Circle the formula on the right for the English sentence on the left

Neither rain nor sleet nor hail will stay these couriers.
R: Rain will stay these couriers.
S:
H:

$\sim R \cdot \sim S \cdot \sim H$

$\sim R \vee \sim S \vee \sim H$

$\sim(R \cdot S \cdot H)$

$\sim(R \vee S \vee H)$

$\sim R \cdot \sim S \cdot \sim H$ You could also circle '$\sim(R \vee S \vee H)$', since it asserts essentially the same thing.

Turn to the next frame.

72

Are you bothered by the absence of *interior* parentheses?

Whenever there is a sequence of conjunctions (or disjunctions), you can omit the parentheses, because the grouping of the conjuncts (or disjuncts) does not affect the truth-value of the compound.

$[(p \cdot q) \cdot r] \cdot s$ has the same truth table as $p \cdot [q \cdot (r \cdot s)]$.

Thus we usually write '$(p \cdot q \cdot r \cdot s)$'.

You may want to *verify* this for the example below.

p	q	r	$[(p \vee q) \vee r]$	$[p \vee (q \vee r)]$
T	T	T		
T	T	F		
T	F	T		
T	F	F		
F	T	T		
F	T	F		
F	F	T		
F	F	F		

Go to the next frame.

73

Truth tables allow us also to determine when two sentences *essentially assert the same thing* in different ways. Two sentences essentially assert the same thing when they are *logically equivalent.* And we have already seen that two sentences are logically equivalent if their truth tables are identical.

Two sentences whose logical forms are p and $\sim\sim p$ are logically equivalent.

p	$\sim p$	$\sim\sim p$
T	F	T
F		

Complete the above truth table and the following one for $\sim(\sim p \mathbin{\&} \sim q)$.

p	q	$\sim p$	$\sim q$	$\sim p \mathbin{\&} \sim q$	$\sim(\sim p \mathbin{\&} \sim q)$
T	T				
T	F				
F	T				
F	F				

What formula, using only one of our basic connectives, is logically equivalent to $\sim(\sim p \mathbin{\&} \sim q)$?

____

Answer in the next frame.

74

$(p \vee q)$ has the same truth table as $\sim(\sim p \ \& \sim q)$.
$(p \ \& \ q)$ has the same truth table as $\sim(\sim p \vee \sim q)$.

These pairs of logical equivalences are known as *De Morgan's Laws*.

De Morgan's Laws are sometimes given as

$\sim(p \vee q)$ is logically equivalent to $(\sim p \ \& \sim q)$.
$\sim(p \ \& \ q)$ is logically equivalent to $(\sim p \vee \sim q)$.

Now you have learned that $(p \rightarrow q)$ is *defined* as $\sim(p \ \& \sim q)$. Using De Morgan's Laws, find a *disjunction* to which $(p \rightarrow q)$ is logically equivalent.

____

$(p \rightarrow q)$ is logically equivalent to $(\sim p \vee \sim\sim q)$, or using your knowledge of double negation, $(\sim p \vee q)$.

Go to the next frame.

75

An easy way to state these equivalences in English is

(1) The *negation of a conjunction* is the *disjunction of the negations* of the components.

(2) The *negation of a disjunction* is the *conjunction of the negations* of the components.

These equivalences, or ones quite similar to them, are known as

____________________ Laws.
(Whose?)

De Morgan's Laws

Consider, for example:
$\sim(B \vee H)$

B: Beer is served.
H: The men are happy.

Interpretation: It is not the case that beer is served or the men are happy.

Using both De Morgan's Laws and simple reflection, you should be able to see that this means the same thing as

____________________________ and ____________________________
(Fill in the English sentences.)

Beer is *not* served *and* the men are *not* happy.

Go to the next frame.

76

Some logical equivalences are trivial, such as

$(p \vee q)$ is logically equivalent to $(q \vee p)$.

Others require some transformation according to remembered logical equivalences, or an appeal to truth tables.

Problem:

$(p \rightarrow q)$ is logically equivalent to which of the following? Use the truth table this time.

$(-p \rightarrow -q)$ $(-q \rightarrow -p)$

p	q	$(p \rightarrow q)$	$-p$	$-q$	$(-p \rightarrow -q)$	$(-q \rightarrow -p)$

____

See the next frame.

77

p	q	$(p \rightarrow q)$	$\sim p$	$\sim q$	$(\sim q \rightarrow \sim p)$
T	T	T	F	F	T
T	F	F	F	T	F
F	T	T	T	F	T
F	F	T	T	T	T

Since $(-q \rightarrow -p)$ is logically equivalent to $(p \rightarrow q)$, we can see how

If cows cannot whistle, then pigs cannot fly

says in a different way:

If pigs can fly, then cows can whistle.

The formula '$(-q \rightarrow -p)$' is often called the *contrapositive* form of the formula '$(p \rightarrow q)$'.

In interpreting symbolic formulas into English sentences, it is a good idea to stick as closely as possible to a literal translation of the symbolic connectives. The awkward and stylistically painful English sentence that usually results can then be rephrased into a more acceptable form.

'$S \rightarrow (L \rightarrow E)$' is more correctly interpreted by which numbered English sentence below? Check one.

S: The switch is on.
L: The lever is thrown.
E: An explosion results.

1. If when the switch is on, the lever is thrown, then an explosion results.

2. If the switch is on, then when the lever is thrown, an explosion results.

Turn to the next frame.

78

Sentence 2, 'If when the switch is on, then when the lever is thrown, an explosion results', or '$S \to (L \to E)$', is logically equivalent to which one of these:

a. $(S \,\&\, L) \to E$

b. $(S \to L) \to E$

p	q	r	$p \,\&\, q$	$p \to q$	$q \to r$	$p \to (q \to r)$	$(p \,\&\, q) \to r$	$(p \to q) \to r$
T	T	T			T			
T	T	F			F			
T	F	T			T			
T	F	F			T			
F	T	T						
F	T	F						
F	F	T						
F	F	F						

____

---

a. $(S \,\&\, L) \to E$ If the columns under '$p \to (q \to r)$' and '$(p \,\&\, q) \to r$' are not identical, find your errors and correct them.

Turn to the next frame.

79

A sentence of the form

$p \rightarrow (q \rightarrow r)$

is logically equivalent to one of the form

$(p \ \& \ q) \rightarrow r$.

What is a correct interpretation of '$(S \ \& \ L) \rightarrow E$'?

1. If the switch is on and the lever is thrown, then an explosion results.

2. The switch is on and if the lever is thrown, then an explosion results.

____

---

1. If the switch is on and the lever is thrown, then an explosion results.

Turn to the next frame.

80

Circle the English sentence that correctly interprets the formula

$A \supset (\sim B \supset C)$

A: Acid and water are mixed.
B: You wish to be burned.
C: You should be careful.

1. If acid and water are mixed and you do not wish to be burned, then you should be careful.

2. When acid and water are mixed, you should be careful if you do not wish to be burned.

Choice 1, turn to frame 81.

Choice 2, turn to frame 82.

81

No! '$A \supset (\sim B \supset C)$' can be partially interpreted as 'If A, then if $\sim B$ then C'.

Furthermore, the latter phrase, 'if $\sim B$ then C', may be rephrased in several ways, such as '$\sim B$ only if C', and 'C, if $\sim B$'.

So one interpretation could be 'If A, then C if $\sim B$'. More explicitly,

> If acid and water are mixed, then you should be careful if you do not wish to be burned.

This sentence corresponds very closely to sentence 2.

Go to the next frame.

82

Sentence 2 is correct. But something puzzling is going on. Let us look at the two answers again.

1. If acid and water are mixed and you do not wish to be burned, then you should be careful.
2. When acid and water are mixed, you should be careful if you do not wish to be burned.

Using the abbreviations given for this example, we symbolize these as

1. $(A \,\&\, {\sim}B) \supset C$
2. $A \supset ({\sim}B \supset C)$

It is easy (but tiresome) to verify that these two sentences are logically equivalent, that they assert "essentially the same thing"!

It will not do to accept as a correct symbolization *any* sentence logically equivalent to the given one, since one can add on to any formula additional formulas that do not affect the truth-values of the original formula. For example, $(p \vee {\sim}p)$ conjoined to any sentence q is logically equivalent to q. Thus, there are an infinite number of structurally different formulas that are logically equivalent to any one formula.

But efforts to restrict the range of acceptable or correct symbolizations of a given sentence have failed to be other than arbitrary.

In symbolizing, therefore, we are guided (intuitively) by the grammatical structure of the original English sentence, and in translating we are again guided by the parentheses and other logical symbols. That is to say, there are no *general* criteria of adequacy for symbolizations or translations, but we must rely on good sense and general agreement.

Go to the next frame.

83

Circle the English sentence that correctly interprets the formula

$(P \vee W) \rightarrow (S \ \& \sim E)$

P: Prices go up.
W: Wages go down.
S: Salaried workers will suffer.
E: Pensioners will suffer.

1. If prices go up, then either wages go down or else salaried workers will not suffer, and furthermore pensioners will suffer.

2. If either prices go up or wages go down, then salaried workers will, and pensioners will not, suffer.

3. Either prices go up, or if wages go down then salaried workers will suffer and it is not the case that pensioners will suffer.

Choice 1, turn to frame 84.

Choice 2, turn to frame 85.

Choice 3, turn to frame 86.

84

You shouldn't have made this error! Sentence 1 begins 'If prices go up, then either. . . .'

This must have come from a formula that begins ('$P \rightarrow$ . . . .'

Return to frame 83 and try again.

85

Sentence 2 is correct!

Determine by truth-table analysis which of the following pairs of formulas are logically equivalent:

1. $\sim p \equiv (\sim p \supset \sim q)$ $\quad$ $p \equiv (p \supset q)$

p	q	
T	T	
T	F	
F	T	
F	F	

2. $\sim(p \vee q)$ $\quad$ $\sim p \;\&\; \sim q$

p	q	

3. $p \rightarrow (q \rightarrow p)$ $\quad$ $\sim p \rightarrow (\sim q \rightarrow \sim p)$

p	q	

The pairs in numbers ______ and ______ are logically equivalent.

If you have	Turn to
1 and 2	frame 87.
1 and 3	frame 89.
2 and 3	frame 90.

86

Sentence 3 is not correct. It comes from the sentence

$$P \vee [W \rightarrow (S \ \& \ {\sim}E)].$$

Watch your parentheses. Return to frame 83.

87

No. Perhaps your truth tables were constructed carelessly. The items in number 2 *are* logically equivalent.

Redo the truth tables for numbers 1 and 3.

1. $\sim p \equiv (\sim p \supset \sim q)$ $\qquad p \equiv (p \supset q)$

p	q	$\sim p$	$\sim q$	$\sim p \supset \sim q$	$\sim p \equiv (\sim p \supset \sim q)$	$p \supset q$	$p \equiv (p \supset q)$
T	T						
T	F						
F	T			F			F
F	F						

compare

Biconditional

3. $p \rightarrow (q \rightarrow p)$ $\qquad \sim p \rightarrow (\sim q \rightarrow \sim p)$

p	q	$q \rightarrow p$	$p \rightarrow (q \rightarrow p)$	$\sim p$	$\sim q$	$\sim q \rightarrow \sim p$	$\sim p \rightarrow (\sim q \rightarrow \sim p)$

Are the items in number 3 logically equivalent? Yes / No

Yes. Turn to frame 90.

No. Turn to frame 88.

88

No. You had better go back over the use of tables to determine logical equivalence.

Review the material in frames 73–79.

Then rework *all* the problems from frame 85 in the space below. Choose the correct answer in frame 85 and follow the directions for that choice.

1. $\sim p \equiv (\sim p \supset \sim q)$ $\qquad$ $p \equiv (p \supset q)$

2. $\sim(p \vee q)$ $\qquad$ $\sim p \ \& \sim q$

3. $p \rightarrow (q \rightarrow p)$ $\qquad$ $\sim p \rightarrow (\sim q \rightarrow \sim p)$

89

Incorrect. The items in number 3 *are* logically equivalent.
Redo the truth tables for numbers 1 and 2.

1. $\sim p \equiv (\sim p \supset \sim q)$ $p \equiv (p \supset q)$

p	q	$\sim p$	$\sim q$	$\sim p \supset \sim q$	$\sim p \equiv (\sim p \supset \sim q)$	$p \supset q$	$p \equiv (p \supset q)$
T	T						
T	F						
F	T			F			F
F	F						

Biconditional

compare

2. $\sim(p \vee q)$ $\sim p \;\&\; \sim q$

p	q	$p \vee q$	$\sim(p \vee q)$	$\sim p$	$\sim q$	$\sim p \;\&\; \sim q$

Are the items in number 2 logically equivalent? Yes / No

Yes. Turn to frame 90.

No. Turn back to frame 88.

90

The items in numbers 2 and 3 are logically equivalent.

2. $\sim(p \vee q)$ $\sim p\ \&\ \sim q$

p	q	$p \vee q$	$\sim(p \vee q)$	$\sim p$	$\sim q$	$\sim p\ \&\ \sim q$
T	T	T	F	F	F	F
T	F	T	F	F	T	F
F	T	T	F	T	F	F
F	F	F	T	T	T	T

3. $p \rightarrow (q \rightarrow p)$ $\sim p \rightarrow (\sim q \rightarrow \sim p)$

p	q	$q \rightarrow p$	$p \rightarrow (q \rightarrow p)$	$\sim p$	$\sim q$	$\sim q \rightarrow \sim p$	$\sim p \rightarrow (\sim q \rightarrow \sim p)$
T	T	T	T	F	F	T	T
T	F	T	T	F	T	F	T
F	T	F	T	T	F	T	T
F	F	T	T	T	T	T	T

Turn to the next page.

Part I Review with Comments

(1) You should be able to select from a list of symbols those symbols that are most commonly used for the five major logical connectives.
Comment: (a) You will often find the symbol for the conditional called the symbol for *material implication.* (Also the biconditional is called *material equivalence.*) These names are traditional but misleading.
(b) In Part II, we shall use just one of each of the symbols you have learned.

(2) You should be able to construct truth tables for each of the five major connectives.
Comment: This is easy if you remember the truth conditions for each connective:
(i) The denial of p is true if p is false; otherwise false.
(ii) The conjunction of p and q is true if both p and q are true; otherwise false.
(iii) The disjunction of p and q is false if both p and q are false; otherwise true.
(iv) The conditional with antecedent p and consequent q is false if p is true and q false; otherwise true.
(v) The biconditional of p and q is true if both p and q are true or both are false; otherwise false.

(3) You should be able to recognize when a symbolic formula correctly exhibits the logical form of an English sentence.
Comment: In English, the kind of connective is usually clear, but the grouping of components can be quite ambiguous. It is necessary to ensure that the parentheses in the formula group together the correct components, and the major connective of the formula is the major connective of the English sentence.
Take, for example, the ambiguous English sentence:
If it rains the picnic is canceled and the club closes at six.
There are two possible symbolizations of this:

$$R \rightarrow (P \ \& \ C) \qquad (R \rightarrow P) \ \& \ C$$

(4) You should be able to reconstruct an English sentence given a symbolic formula and the sentence abbreviations.
Comment: If worse comes to worst you could scratch out the connective symbols and write in 'and', 'if . . . , then', and so on.

In polishing up the style of the resulting English sentence, be careful that you do not change the truth conditions.
An example of how one might go *wrong* is '$\neg(Q \,\&\, R)$'.

Q: The queen arrives.
R: John remains seated.

(i) It is not the case Q and R.
(ii) The queen does not arrive and John remains seated. WRONG.

The correct interpretation is, as you know, 'It is not the case that *both* the queen arrives and John remains seated'.

(5) You should be able to determine when two formulas are logically equivalent.
Comment: Sentence logic is a simple system to work with because it is so easy to determine if (and if not) two formulas are logically equivalent. To determine this, you see if the truth tables for the two formulas are identical. This can *always* be done.
For example, are $[p \rightarrow (q \rightarrow p)]$ and $[q \rightarrow (p \rightarrow q)]$ logically equivalent? Draw up a truth table and see.

p	q	$p \rightarrow q$	$q \rightarrow p$	$p \rightarrow (q \rightarrow p)$	$q \rightarrow (p \rightarrow q)$
T	T				
T	F				
F	T				
F	F				

With experience you will recognize that one formula can be *transformed* into another by substituting some common logical equivalences, such as

Double Negation	$\sim\sim p$	p
De Morgan's Laws	$\sim(p \vee q)$	$\sim p \,\&\, \sim q$
	$\sim(p \,\&\, q)$	$\sim p \vee \sim q$
Contraposition	$p \rightarrow q$	$\sim q \rightarrow \sim p$
Exportation	$(p \,\&\, q) \rightarrow r$	$p \rightarrow (q \rightarrow r)$

Thus one might use a sequence of transformations, each formula of the sequence being logically equivalent to the preceding one, and hence each being equivalent to any of the others.

$p \supset (-q \vee r)$

$-(-q \vee r) \supset -p$	by Contraposition
$(--q \cdot -r) \supset -p$	De Morgan
$(q \cdot -r) \supset -p$	Double Negation
$(-r \cdot q) \supset -p$	(Commutativity)
$-r \supset (q \supset -p)$	Exportation
$-r \supset -(q \cdot --p)$	(Definition)
$-r \supset -(q \cdot p)$	Double Negation
$(q \cdot p) \supset r$	Contraposition
$(p \cdot q) \supset r$	(Commutativity)

All of these are equivalent, but to show the first equivalent to the last, you could use truth tables to settle the issue. You can always use truth tables to determine logical equivalence.

Self-Examination

1. Match the proper symbols with the indicated truth-functional connectives. Not all symbols will be used, and in some cases, more than one symbol should be matched with a connective.
   SYMBOLS: $\star \quad \rightarrow \quad \vee \quad \square \quad \equiv \quad \cdot \quad / \quad \supset \quad \leftrightarrow \quad \triangle \quad \& \quad + \quad \sim \quad \wedge$

   Biconditional ______________

   Conditional ______________

   Disjunction ______________

   Conjunction ______________

   Negation ______________

2. Write in the name of the truth-functional connective whose truth conditions are given below:
   A. False when both components are false; true otherwise.

      ____

   B. True when its component is false; false otherwise.

      ____

   C. True when both components are either both true or both false; false otherwise. ______________
   D. True when both components are true; false otherwise.

      ____

   E. False when the first component is true and the second false; true otherwise. ______________

3. Select the formula that correctly symbolizes the English sentence 'Either the army is strong and the treasury full, or the people are restless'.
   A: The army is strong.
   T: The treasury is full.
   P: The people are restless.
   (a) $(A \vee T) \wedge P$
   (b) $(A \;\&\; T) \vee P$
   (c) $A \rightarrow (T \rightarrow P)$

(continued)

(d) $(A \equiv T) \supset P$
(e) $A \vee (T \cdot P)$

4. Select the English sentence that correctly interprets the formula '$(H \rightarrow M) \cdot (\sim H \rightarrow G)$'.
   H: I am honest.
   M: Men hate me.
   G: The gods hate me.

   (a) If I am honest, then men hate me, and if I am not honest, the gods hate me.
   (b) If I am honest and men hate me, then it definitely is the case that the gods do not hate me.
   (c) Whether I am honest or not, either men hate me or the gods do.
   (d) Either men hate me or the gods hate me, and I am either honest or not honest.

5. Using truth tables, determine which formula is logically equivalent to $[p \vee (q \& \sim p)]$.

   (a) $(p \vee q) \& \sim p$
   (b) $(p \vee q)$
   (c) $p \& (q \vee \sim p)$
   (d) $(\sim p \& q) \supset p$

ANSWERS

1. Biconditional $\equiv$ $\leftrightarrow$

   Conditional $\rightarrow$ $\supset$

   Disjunction $\vee$

   Conjunction $\cdot$ & $\wedge$

   Negation $\sim$

2.
   A. Disjunction
   B. Negation
   C. Biconditional
   D. Conjunction
   E. Conditional

3. (b) 4. (a) 5. (b)

End of Part I

Part 2

Predicate Logic with Identity (PLI)

Introduction

The justification for much of our reasoning depends not only on the structure of compound sentences but also on the internal structure of simple sentences, as the examples in the text will show. We need therefore to enrich the artificial language of Part I so that we can represent the logically relevant internal structures of many important natural-language sentences.

Building on what we learned in Part I, we will arrive at an artificial language (PLI) that will enable us to refer to things and their properties and relations. We are faced, however, with the same problem as before—namely, that of having a natural language (English) on the one hand, and an artificial language (PLI) on the other. After learning what PLI is, then, you will learn how to recognize when a formula of PLI correctly symbolizes an English sentence, and conversely, when an English sentence correctly interprets a formula of PLI.

In sentence logic (SL), one simple sentence might be '*A*' and another '*B*.' These would not be logically equivalent in SL. But finer analysis of the internal structure of these sentences could reveal that they were logically equivalent in PLI. You will learn to recognize some elementary logical equivalences between formulas of PLI, as well as some elementary contradictories of some formulas of PLI.

To summarize, upon completion of Part II you should be able to

(1) Select the symbolic formula of PLI that correctly symbolizes an English sentence of moderate complexity.
(2) Select an English sentence that correctly interprets a given formula of PLI of moderate complexity.
(3) Recognize some elementary logical equivalences between formulas of PLI.
(4) Recognize some elementary contradictories of given formulas of PLI.
(5) Interpret an expression containing the abstraction operator.

TEAR THIS SHEET OUT

You may use this sheet as your shield. You may also do scratch work on it.

DO NOT MAKE NOTES.

-------------------------- fold here --------------------------

91

While it is basic and fundamental, the analysis of truth-functional compound sentences is usually not fine enough for our purposes.

Herman is a dog

and

Rover is a dog

are clearly similar sentences, yet this similarity is lost when we abbreviate them '*H*' and '*R*' respectively.

Herman is small

and

Herman is brown

are also sentences whose similarity is lost when we use '*S*' and '*B*' respectively as abbreviations.

All four of our example sentences are simple, but what is the major similarity between the sentences '*S*' and '*B*' that is not also shared by the sentences '*H*' and '*R*'?

____

____

(Write your answer.)

'*B*' and '*S*' have the same *subject,* Herman.

Go to the next frame.

92

We shall abbreviate the *subject terms* in a sentence with *lower-case letters* from the beginning of the alphabet. Therefore

a, *b*, *c*, . . ., *l*, *m*

are *individual constants*. So 'Herman is brown' becomes '*h* is brown', and 'Herman is small' becomes '*h* is small'.

Work out the unfinished examples below:

1. Rover is a dog.
   e: Rover — *e* is a dog.

2. Charles did not steal the Rembrandt from the Louvre.
   c: Charles
   b: the Rembrandt
   l: the Louvre — *c* did not steal *b* from *l*.

3. Barbara is a tall, thin blond.
   b: Barbara — ______________

---

3. *b* is a tall, thin blond.

---

4. Fielding wrote *Tom Jones* but not *Moll Flanders*.
   f: Fielding
   j: *Tom Jones*
   m: *Moll Flanders* — ______________

---

4. *f* wrote *j* but not *m*.

---

5. If Tom and Jerry are elected, then Jane has done her work well.
   m: Tom
   j: Jerry
   a: Jane — ______________

---

5. If *m* and *j* are elected, then *a* has done her work well.

Go to the next frame.

93

1. (a) Fielding wrote *Tom Jones* but not *Moll Flanders*.
   (b) Fielding wrote *Tom Jones* but <u>Fielding</u> did not write *Moll Flanders*.

2. (a) Barbara is a blond or else she has dyed her hair again.
   (b) Barbara is a blond or else <u>Barbara</u> has dyed her hair again.

3. (a) Romeo and Juliet love each other.
   (b) Romeo loves Juliet and Juliet loves Romeo.

Make the subject (or subjects) explicit in the remaining examples.

4. (a) If Tom is drinking champagne, then he has won the race.
   (b) ______________________________

   ____

---

4. (b) If Tom is drinking champagne, then Tom has won the race.

---

5. (a) Either Carol and Lois have both passed, or they have failed.
   (b) ______________________________

   ____

---

5. (b) Either Carol and Lois have both passed, or Carol and Lois have failed.

---

6. (a) The Van Gogh has been repaired but not restored.
   (b) ______________________________

   ____

---

6. (b) The Van Gogh has been repaired, but the Van Gogh has not been restored.

Go to the next frame.

94

Once a sentence has been rephrased so as to make the subjects explicit, you can then replace the names by the *individual constants* which are their abbreviations. Thus

Fielding wrote *Tom Jones* but not *Moll Flanders*

would look like

f wrote *j* but *f* did not write *m*.

What would these look like?

1. Romeo and Juliet love each other.
   e: Romeo
   j: Juliet

   ___

---

1. *e* loves *j* and *j* loves *e*.

---

2. Either Cheryl and Carol have decided not to come, or they have missed the plane.
   h: Cheryl
   c: Carol

   ___

---

2. Either *h* and *c* have decided not to come, or *h* and *c* have missed the plane.

Go to the next frame.

95

Let's try to rephrase sentences in such a way that we do not have multiple subjects. That is, let us try to have one subject for each *predicate*.

Hence let us rewrite 'h and c have missed the plane' as 'h has missed the plane and c has missed the plane'.

Some examples:

1. (a) Either d or j was elected.
   (b) Either d was elected or j was elected.

2. (a) c selected e, f, and g.

   (b) ______________________________

   ____

---

2. (b) c selected e and c selected f and c selected g.

---

3. (a) k and h are each harder than m.

   (b) ______________________________

   ____

---

3. (b) k is harder than m and h is harder than m.

Go to the next frame.

96

These examples have been easy, but you must exercise caution in splitting up subjects. Sometimes the sense of the *predicate* demands that the subjects be kept together.

k and *b* are twins

should *not* be written

k is a twin and *b* is a twin.

Also, if *j* and *c* *each* bought a dress, we would write:

j bought a dress and *c* bought a dress.

But if they pooled their money and bought one dress for both of them, we would write:

j and *c* (jointly) bought a dress.

Circle the sentence where the subjects very definitely ought *not* to be split up:

1. *h* and *k* joined the party later.

2. *d* and *g* were tied for second prize.

3. *m*, *k*, and *i* are famous scientists.

4. *j* stole *e* and *f*.

---

2. *d* and *g* were tied for second prize.

Go to the next frame.

97

You will recall that we said 'Herman is a dog' and 'Rover is a dog' were similar sentences. One important similarity between these sentences is that they *assert the same thing about different individuals.*

Sentences	Predicates
1. Herman is a dog.	______ is a dog.
2. Romeo loves Juliet.	______ loves ______.
3. Jim gave the pin to Mary.	______ gave ______ to ______.

How do we obtain the *predicate* of a simple sentence like those above?

____

____

(Write out the answer.)

(In your own words) Delete the subject terms or names from the sentence.

Go to the next frame.

98

Another way of looking at the matter is this. When all the *blanks* in a predicate are filled with *names*, we obtain a *sentence*.

So when the predicate '______ is a dog' has a name placed in the blank, we have a sentence. If the name is 'Fido' (abbreviated '*f*'), we have the *true* sentence '*f* is a dog', and if the name is 'George Washington' (abbreviated '*g*'), we have the *false* sentence '*g* is a dog'.

______ is a small, brown, noisy dog

can be rephrased as

______ is small and ______ is brown and ______ is noisy and

______ is a dog.

Some of the eight predicates below are *complex* and can be resolved into a sequence of simpler predicates joined by such connectives as 'or' or 'and'. Circle the *complex* predicates.

1. ______ struck ______.
2. Tall, lean ______ struck short, fat ______.
3. ______ has blue eyes and blond hair.
4. ______ has a peculiar nose.
5. ______ is much greater than ______.
6. ______ has more courage or less brains than ______.
7. ______ is between ______ and ______.
8. ______ very nearly matches ______ in color.

---

2. Tall, lean ______ struck short, fat ______.
3. ______ has blue eyes and blond hair.
6. ______ has more courage or less brains than ______.

If you are correct, go to frame 100.

If you are incorrect, turn to the next frame.

99

Let us look at some of the examples more closely. Example 2, 'Tall, lean ______ struck short, fat ______', can be rephrased as '______ is tall and ______ is lean and ______ struck ______ and ______ is short and ______ is fat'.

Example 4, '______ has a peculiar nose', *cannot* be rephrased as '______ is peculiar and ______ has a nose'. We can admit that further aesthetic analysis may reveal those properties that make it a peculiar nose, yet *logically*, and with no more context specified, having a peculiar nose is a simple property.

Example 6, '______ has more courage or less brains than ______', can be analyzed into '______ has more courage than ______ or ______ has less brains than ______'.

Example 7, '______ is between ______ and ______', is as simple a predicate as one can find. No one thing, or pair of things, is claimed to have a property or to exemplify a relation. Only a triplet of individuals could exemplify the simple relation denoted by the simple predicate.

Go to the next frame.

100

Before we go on, you may wish to review the things you have learned.

Subjects are sometimes implicit and unnamed, and sometimes referred to by pronouns. You should know how to rewrite a sentence to make explicit all subject terms.

Make the subject terms explicit in the following sentences:

1. John hit the ball and trotted to first base.

---

1. John hit the ball and *John* trotted to first base.

---

2. If he knows what he is doing, Bill is unlikely to receive a shock.

---

2. If *Bill* knows what *Bill* is doing, Bill is unlikely to receive a shock.

If you are unclear or have made a mistake, review the material in frame 93.

Otherwise, turn to the next frame.

101

Explicit subject terms can be replaced by abbreviations (lower-case letters from the beginning of the alphabet called *individual constants*).

But some sentences contain *multiple* subject terms and some contain *complex* predicates. We usually must rephrase such sentences. Simplify and use abbreviations in the following examples:

1. Herman and Fido are brown dogs.
   h: Herman
   f: Fido
   h is brown and *h* is a dog and *f* is brown and *f* is a dog.

2. Either Betty or Carol spilled the milk and did not wipe it up.
   b: Betty
   c: Carol
   m: the milk

---

2. (Either) *b* spilled *m* and *b* did not wipe *m* up, or *c* spilled *m* and *c* did not wipe *m* up.

---

3. Ken is conceited and egotistical, unless Henry is sadly mistaken about him.
   k: Ken
   h: Henry

---

3. *k* is conceited and *k* is egotistical, unless *h* is sadly mistaken about *k*.

Go to the next frame.

102

We are now going to put capital letters to a new use. All the capital letters except '*F*', '*G*', and '*H*' will be used to abbreviate predicates.

Returning to our good friend Fido, if

D: ______ is a dog
f: Fido

then the abbreviation of the sentence 'Fido is a dog' is '*Df*'.

1. Fido is faithful. ______*Af*______
   A: ______ is faithful.

2. Herman is a dog. ______*Dh*______
   h: Herman

3. Herman is a faithful dog. ______*Dh* and *Ah*______

4. If Green did not commit the murder, then he is absolutely innocent.
   C: ______ committed the murder. ______If not *Cg* then *Ig*______
   I: ______ is absolutely innocent.
   g: Green

Complete the examples:

5. If Charles is foolhardy, he is not a hero.
   L: ______ is foolhardy. ______________
   E: ______ is a hero.
   c: Charles

---

5. If *Lc*, not *Ec*

(continued on next page)

6. Alice and Betty are tall blonds. ________________
   a: Alice
   b: Betty
   T: ______ is tall.
   B: ______ is a blond.

---

6. Ta and Ba and Tb and Bb (Any order of these four is correct—for example, Ta and Tb and Ba and Bb.)

If you wish more discussion of these examples, go to frame 103.

Otherwise, go to frame 104.

103

In the example

If Charles is foolhardy, he is not a hero

we should first replace the pronoun with the name of the individual referred to. Thus:

If Charles is foolhardy, Charles is not a hero.

We can now abbreviate names with some individual constants: 'If *c* is foolhardy, *c* is not a hero'. '*c* is foolhardy' is abbreviated '*Lc*', and '*c* is a hero' is abbreviated '*Ec*'. The denial of '*Ec*' is 'not *Ec*', and so we have

If *Lc*, not *Ec* .

The example

Alice and Betty are tall blonds

exemplifies both a multiple subject and a complex predicate. First we note that the same (complex) property—the property of being a tall blond—is predicated of both Alice and Betty. Abbreviating their names, we write:

a is a tall blond and *b* is a tall blond.

Decomposing the complex predicate, '______ is a tall blond', we write:

a is tall and *a* is a blond and *b* is tall and *b* is a blond.

Now we abbreviate these predicates, obtaining

Ta and *Ba* and *Tb* and *Bb*.

Go to the next frame.

104

As long as we deal only in simple properties involving only simple subjects, we could get along with the abbreviatory devices we have been using—for example, '*D*' abbreviates '______ is a dog'.

But when we turn to *relations* obtaining between two or more subjects (formally, predicates with two or more blanks), the situation is different. The abbreviations of predicates must show the *number* and *order* of the blanks in the original predicate.

After all, two *is* less than twelve, and we should make a serious mistake if we changed the order of the numerals in the symbolization of the sentence 'Two is less than twelve'.

1. John loves Mary, but Mary doesn't love him.
   L __ __: ______ loves ______.
   j: John
   m: Mary

   Ljm but not Lmj

2. The blue car is between the red car and the truck.
   B __ __ __: ______ is between ______ and ______.
   b: the blue car
   d: the red car
   k: the truck

   Bbdk

Using the above abbreviations, symbolize the following sentences:

3. John loves Mary and Carol.
   c: Carol.

   ____

---

3. *Ljm* and *Ljc*

---

4. The truck is between the red car and the blue car.

   ____

---

4. *Bkdb*

Go to the next frame.

105

The blanks in the abbreviated predicate (the capital letter) correspond, in specific order, to the blanks in the *unabbreviated predicate*. When symbolizing you should take care to have the *abbreviated names* in the correct order, and when interpreting you should take care that the actual names are in the right order when written in the blanks of the predicate.

The symbolic formula '*Cmj*' can be interpreted by which of the following English sentences?

C ___ ___: ______ called ______ .

m: Mike

j: Jane

1. Jane was called by Mike.
2. Mike was called by Jane.
3. Jane called Mike.
4. Mike called Jane.

____

'*Cmj*' may be interpreted as number 4, 'Mike called Jane'.

Go to the next frame.

106

The presence of words such as 'but', 'and', 'or', 'not', and so forth should have suggested to you what the next and final step in symbolizing is. We have been conjoining and disjoining and denying sentences all along. So now we'll use the symbols from Part I.

1. Fielding wrote *Tom Jones* but not *Moll Flanders*.

   W __ __: ______ wrote ______ .

   f: Fielding
   j: *Tom Jones*
   m: *Moll Flanders*

   (Wfj but not Wfm)
   (Wfj & $\sim Wfm$)

2. ($Lg \rightarrow \sim Eg$)

   L __: ______ is foolhardy.
   E __: ______ is a hero.
   g: George

   If George is foolhardy, he is not a hero.

3. Lean Cassius stabbed Julius in the stomach or the chest.

   L __: __ is lean.
   S __ __ __: __ stabbed __ in __ .
   c: Cassius
   j: Julius
   m: Julius's stomach
   h: Julius's chest

   Lc & ($Scjm \vee Scjh$)

   (A semi-literal rendition of this formula might be 'Cassius is lean and either he stabbed Julius in the stomach or he stabbed Julius in the chest'.)

4. (Sb & Rb & Db & Bb)

   S __: ______ is small.
   R __: ______ is red.
   D __: ______ is a dog.
   B __: ______ barks at strangers.
   b: Bowser

   ____

   (Render into decent English.)

---

4. A good English interpretation is 'Bowser is a small red dog who barks at strangers'.

If you have nothing like the correct answer or are somewhat weak on this material, turn to the next frame.

If you are correct and wish to go on, turn to frame 108.

107

We are attempting to rephrase complicated sentences as truth-functional compounds of *elementary sentences.* Elementary sentences, so far, are one-place predicates with one name in the blank, or two-place predicates with two names in the blanks, and so on. Elementary sentences are abbreviated by one-place predicate letters followed by one individual constant, two-place predicate letters followed by two individual constants, and so on.

When we approach a sentence such as

> If Bill is guilty, he will be fined or given a warning by the judge

we first make the subject terms explicit in every instance:

> If Bill is guilty, Bill will be fined by the judge or Bill will be given a warning by the judge.

As far as the *sentential* symbolism is concerned, this has the form

$$p \rightarrow (q \vee r).$$

We now want to analyze the sentences p, q, and r into their subject-predicate form.

The antecedent of the conditional is 'Bill is guilty', which is symbolized (with obvious abbreviations) 'Lb'. (We are saving 'F', 'G', and 'H'.)

The consequent of the conditional is a disjunction, and the left-hand side (the left disjunct) is 'Bill will be fined by the judge'. Let 'N . . . ______ ' mean that the first-named individual is fined by the second individual—that is,

> . . . is fined by ______ .

Then 'Bill is fined by the judge' is symbolized 'Nbj'.

And by the same sort of passive construction, the right-hand disjunct is symbolized 'Wbj'.

Our completed symbolization looks like this:

$$Lb \rightarrow (Nbj \vee Wbj).$$

Go to the next frame.

108

Circle the formula symbolizing the given English sentence.

1. If John is taller than Mary, then if Mary is taller than Ann, John is taller than Ann.

   T ___ . . . : ______ is taller than . . .
   j: John
   m: Mary
   a: Ann

   $Tj \rightarrow (Tm \rightarrow Ta)$

   $Tjm \rightarrow (Tma \rightarrow Tja)$

   $Tjm \rightarrow (Tam \;\&\; Tja)$

   $Tmj \rightarrow (Tma \rightarrow Tja)$

---

1. $Tjm \rightarrow (Tma \rightarrow Tja)$

---

2. Either Jim stole the necklace, or he hasn't an alibi.

   S . . . ___: . . . stole ______ .
   A ___: ______ has an alibi.
   j: Jim
   e: the necklace

   $Sje \vee {\sim}Ah$

   $Sje \vee Aj$

   $Sej \vee Aj$

   $Sje \vee {\sim}Aj$

---

2. $Sje \vee {\sim}Aj$

---

3. Lean, hungry Cassius stabbed fat Julius.

   L ___: ______ is lean.
   U ___: ______ is hungry.
   A ___: ______ is fat.
   c: Cassius
   j: Julius
   S ___ . . . : ______ stabbed . . .

   $S(Lc \;\&\; Uc), Aj$

   $Lc \;\&\; Uc\ S\ Aj$

   $Lc \;\&\; Uc \;\&\; Aj \;\&\; Scj$

   $(Lc \;\&\; Uc) \rightarrow (Scj \;\&\; Aj)$

---

3. $Lc \;\&\; Uc \;\&\; Aj \;\&\; Scj$

Go to the next frame.

109

Circle the English sentence that interprets the given formula.

$Kga \rightarrow (Pa \& {\sim}Am)$

K __ . . . : ______ kissed . . .
P __: ______ is pleased.
A __: ______ is happy.
g: George
a: Alice
m: Mary

1. Alice is pleased and Mary is not happy, if George kissed Alice.

2. If George kissed Alice, then Alice is pleased only if Mary is unhappy.

---

1. Alice is pleased and Mary is not happy, if George kissed Alice.

$Tbb \& {\sim}Sbd$

T __ . . . : ______ talks to . . .
S __ . . . : ______ speaks to . . .
b: Bill
d: Dr. Nertz

1. Bill talks and he says nothing to Dr. Nertz.

2. Bill talks to Bill but refuses to speak to Dr. Nertz.

3. Bill talks to himself, but he does not speak to Dr. Nertz.

---

3. Bill talks to himself, but he does not speak to Dr. Nertz.

Go to the next page.

TEAR THIS SHEET OUT

Individual constants: a, b, c, . . . , m
Predicates: A, B, C, D, E, I, J, K, . . .

-------------------------- fold here --------------------------

110

When we fill in the blanks of a predicate with names of individuals we obtain a sentence. But many sentences have *no* names occurring in them. Consider, for example,

> There are some dogs that do not bark.
> All human beings are mortal.
> If anyone falls asleep, he will answer to no one.

To symbolize such sentences we need to use a device (attributed to Bertrand Russell) commonly called a *sentential function* (or propositional function).

Using *individual variables*

$u, v, w, x, y, z,$

we *construct sentential functions from predicates.*

PREDICATE	SENTENTIAL FUNCTION
_______ is a dog.	x is a dog.
_______ answers to . . .	x answers to y.
_______ is a human being.	x is a human being.

Choose from the right-hand column the appropriate term for the item on the left.

1. Rover is a dog.	____________________	Individual name
2. m	____________________	Individual variable
3. y is a dog.	____________________	Sentence
4. Herman	____________________	Predicate
5. _______ is a dog.	____________________	Sentential function
6. x	____________________	Individual constant

---

1. Sentence
2. Individual constant (frame 91)
3. Sentential function
4. Individual name
5. Predicate (frames 97, 104)
6. Individual variable

Go to the next frame.

111

The abbreviation of a sentential function is, as you would expect, the abbreviation of the predicate followed by the appropriate individual variables.

SENTENTIAL FUNCTION	SYMBOLIZED
x is a dog.	Dx
x answers to y.	Axy
x is a human being.	Bx

You will find some authors writing a two-place predicate symbol *between* the two terms belonging to it. For instance, in mathematics you will find '$x < y$' for 'x is less than y'. Later we shall occasionally adopt this way of symbolizing, but for the present we shall write '$L\ x\ y$' for 'x is less than y'. Writing the predicate symbol between the terms is awkward when more than two terms are involved.

'x is between y and z' in our notation is '$Bxyz$', but some authors are forced to write '$xBy;z$' or '$yBz(x)$' or some other confusing symbolization.

One way to construct a *sentence* from a sentential *function* is to replace the individual v_______s by individual n_____s.
(Fill in the blanks.)

---

variables; names.

Go to the next frame.

112

Suppose we write:

For some individual x, x is a dog and x barks.

This is symbolized, thus far, as

For some individual x, $(Dx \ \& \ Bx)$.

What shall we call such an expression? Noting that what we have written makes an assertion, or in other words, is either true or false, we conclude that this is a *sentence*.

If there were *no* dogs that barked, all the following sentences would be *false*:

(1) Some dogs bark.
(2) There are dogs that bark.
(3) There are some dogs that bark.
(4) There is at least one dog that barks.

In spite of the fact that the first three of these sentences *suggest* that there exist more than one barking dog, we shall understand them in the minimal sense of (4): *There is at least one x such that* x is a dog and x barks.

Consider these abbreviations:

j: Johnny
Lxy: x loves y.

Complete example 2 in the same way as example 1.

1. Johnny loves someone.
   (a) j loves someone.
   (b) There is an x such that j loves x.
   (c) There is an x such that Ljx.

2. Someone loves Johnny.
   (a) Someone loves j.
   (b) There is an x such that ______________________
   (c) ______________________

---

2. (b) There is an x such that x loves j.
2. (c) There is an x such that Lxj.

Go to the next frame.

113

To symbolize the concept of there *existing* at least one thing that satisfies a sentential function, we write a backwards 'E' with the variable. Thus

$(\exists x)(Dx \ \& \ Bx)$

symbolizes 'There is at least one x such that x is a dog and x barks'.

$(\exists x)(Ljx)$: Johnny loves someone.
$(\exists x)(Lxj)$: Someone loves Johnny.

Since the variables merely keep track of the blanks in a predicate, which variable you use is not important. '$(\exists x)(Dx \ \& \ Bx)$' means exactly the same thing as '$(\exists y)(Dy \ \& \ By)$'.

Using 'y' for the sake of variety, symbolize completely:

Some sailors do *not* drink whiskey.
Sy: y is a sailor.
Dy: y drinks whiskey.

1. For some y, y is a sailor and ______________________________

2. ____

---

1. For some y, y is a sailor and y does not drink whiskey.
2. $(\exists y)(Sy \ \& \sim Dy)$
   (Don't forget parentheses.)

Go to the next frame.

114

A sentential function is *existentially quantified* if it is preceded by an *existential quantifier*, '(∃)', with an individual variable.

Just as our understanding of the connectives was made precise and complete through the study of truth conditions, so here too we must examine the truth conditions of an existential sentence. We have been saving '*F*', '*G*', and '*H*' for this discussion.

'*Fx*' will represent for us any arbitrary sentential function of x; '*Fy*' is that same sentential function with '*y*' in place of '*x*'; '*Fa*' is a sentence formed from that same sentential function with '*a*' filling the blanks of the predicates occurring in *F*. Speaking generally, '$(\exists x)Fx$' is representative of any existentially quantified sentential function of x.

'*Fx*' is representative of all of the following sentential functions *except one*. Which one is *not* represented by '*Fx*'?

1. $(Sx \,\&\, {\sim}Dx)$

2. Lxj

3. $(Dy \,\&\, By)$

4. $[Ax \,\&\, (Lxm \vee Lxk)]$

---

3. '$(Dy \,\&\, By)$' is *not* represented by '*Fx*', since it is a sentential function of y. It would be represented by '*Fy*'.

Go to the next frame.

115

If we have a sentence $(\exists x) Fx$, under what conditions is it true? And under what conditions is it false?

These questions can be answered in a simple manner satisfactory for our purposes in this program.

$(\exists x)Fx$ is *true* if there is *at least one* name, or denoting expression, which when substituted for 'x' in Fx yields a true sentence.

$(\exists x)Fx$ is *false* if *every* name and denoting expression when substituted for 'x' in Fx yields a false sentence.

Suppose that, as a matter of fact, John loves Mary but not Helen. Then (for Lxy abbreviating x loves y, and obvious abbreviations of John, Mary, and Helen)

Ljm is true, and Ljh is false.

What is the truth-value of $(\exists x)Ljx$? (Interpreted: John loves someone.)

True / False ____

True. Go to frame 116.

False. Go to frame 117.

116

'($\exists x$)Ljx' is true, since at least one of the substitution instances of 'Ljx' (namely 'Ljm') is true.

What if *every* substitution instance of 'Ljx' were true? That is, what if 'Ljm' and 'Ljh' are true and similarly for every possible substitution instance—including the case where 'Ljj' is true? That would mean that John loves *everything* (including himself).

When *each and every* substitution instance of a sentential function Fx is true, we shall say the *universal quantification* of Fx is true. The universal quantification is symbolized '$(x)Fx$'.

(Some authors use an inverted 'A'—'$(\forall x)Fx$'—where the upside-down 'A' comes from '*a*ll' just as '∃' came from '*e*xists'.)

$(x)Fx$ is true if every substitution instance of Fx is true.

$(x)Fx$ is false if there is at least one ______________________________

____

(Complete the statement.)

$(x)Fx$ is false if there is at least one substitution instance of Fx that is false.

Go to frame 118.

117

You said 'John loves someone' is false. But one of our assumptions was that John loves Mary! Remember that the sentence $(\exists x)Fx$ needs to have only *one* substitution instance of Fx true in order for the whole sentence $(\exists x)Fx$ to be true. The fact that John does *not* love Helen is really irrelevant.

Review frame 115 and choose the correct answer.

118

'$(x)Fx$' can be read a number of ways:

For all x, Fx
Each x is such that Fx
For any arbitrary x, Fx
Every x is such that Fx

It is true that all Greeks are mortal, yet it would be incorrect to symbolize the sentence

All Greeks are mortal
Rx: x is a Greek
Mx: x is mortal

as '$(x)(Rx \& Mx)$'.

Why is this incorrect?

Because not every substitution instance ______________________

____

____

(Complete.)

Because not every substitution instance of '$(Rx \& Mx)$' is a true sentence. For instance, substitution of 'Napoleon' for 'x' yields 'Napoleon is a Greek *and* he was mortal'.

Go to the next frame.

119

$(x)Fx$ is true when there are no false substitution instances of Fx.

How are we to symbolize 'All Greeks are mortal'?

Consider under what conditions we would take this sentence to be true. As we go through the entire universe, picking and choosing individuals, *when we find a Greek, he turns out to be mortal.*

We can restate 'All Greeks are mortal' as

For *any* individual x, if x is a Greek, then x is mortal.

What is the symbolization of the rephrased sentence?

Rx: x is a Greek.
Mx: x is mortal.

1. $(\exists x)(Rx \;\&\; Mx)$

2. $(x)(Rx \;\&\; Mx)$

3. $[(x)Rx \rightarrow Mx]$

4. $(x)(Rx \rightarrow Mx)$

Choice 1, turn to frame 120.

2, turn to frame 121.

3, turn to frame 122.

4, turn to frame 123.

120

'$(\exists x)(Rx \;\&\; Mx)$' is interpreted as 'There is at least one Greek who is mortal'.

'$(\exists x)$' is an *existential* quantifier of 'x'. Our problem, however, deals with a *universally* quantified sentence.

Review frame 119 and choose another answer.

121

We have just finished a discussion showing this can't be the correct answer.

Reread the material on frame 118. Continue the program from there.

122

In symbolizing

For any x, if x is Greek, then x is mortal

we might write, as a first step,

For any x, if Rx then Mx.

Notice that the sentential function that is being quantified by 'For any x' is 'if Rx then Mx', which is symbolized '$(Rx \rightarrow Mx)$'.

Writing in the universal quantification of 'x', we obtain choice 4:

$(x)(Rx \rightarrow Mx)$ (Note the parentheses!)

In choice 3, '$[(x)Rx \rightarrow Mx]$', the sentential function quantified by '(x)' is just 'Rx'. Thus we have a conditional *formula* whose *antecedent* is '$(x)Rx$'—which is the symbolization of the complete sentence 'Everything is Greek'.

So choice 3 symbolizes 'If *everything is Greek*, then x is mortal'. This expression is not a sentence because 'x' is not within the *scope* of the quantifier and one cannot tell if the whole expression is true or false.

In order to show that the individual chosen in the antecedent is the same individual to be chosen in the consequent, we must enclose the whole expression in parentheses or brackets before writing the initial quantifier. That makes all free occurrences of the variable within these parentheses fall within the scope of the quantifier.

The scope of a quantifier is the formula occurring immediately to the right of the quantifier:

$(x)(Rx \rightarrow Mx)$ $(x)Rx \rightarrow Mx$

Go to frame 124.

123

'$(x)(Rx \rightarrow Mx)$' is the correct symbolization of

For any x, *if* x is a Greek, *then* x is mortal.

Now if you are quite certain why '$[(x)Rx \rightarrow Mx]$' (choice 3) is incorrect, go on to frame 124.

But if you are puzzled about the position of the left parenthesis and bracket, turn back to frame 122 for a discussion about the *scope* of a quantifier.

124

Remembering that for any sentential function Fx, the sentence

$(\exists x)Fx$ is true if *some* substitution instance of Fx is true

and

$(x)Fx$ is true if *every* substitution instance of Fx is true,

we can symbolize quite easily a great many English sentences. First, some fairly uncomplicated ones:

1. All ravens are black.
   Rx: x is a raven.
   Bx: x is black.

   $(x)(Rx \rightarrow Bx)$

2. Horses and cows are mammals.
   Ox: x is a horse.
   Cx: x is a cow.
   Mx: x is a mammal.

   (For any x, if x is either a horse or cow, then x is a mammal.)

   $(x)[(Ox \vee Cx) \rightarrow Mx]$

Another rephrasing gives 'For any x, if x is a horse, then x is a mammal, *and*, for any x, if x is a cow, then x is a mammal'. So another symbolization of sentence 2 is

$(x)(Ox \rightarrow Mx)$ & $(x)(Cx \rightarrow Mx)$

Why is '$(x)[(Ox$ & $Cx) \rightarrow Mx]$' an incorrect symbolization of 'Horses and cows are mammals'?

Because ____

____

____

Go to the next frame.

125

'Horses and cows are mammals' should not be symbolized as '$(x)[(Ox \& Cx) \rightarrow Mx]$' because no individual is both a horse *and* a cow. Therefore, *instances* of '$(Ox \& Cx) \rightarrow Mx$' will never have a true antecedent. The original sentence does *not* assert that anything that is *both* a horse and a cow is a mammal; rather that anything that is *either* a horse or cow is a mammal.

Another example is

Every voter is a Republican or a Democrat.

Vx: x is a voter.
Rx: x is a Republican.
Dx: x is a Democrat.

Choose the correct symbolization.

1. $(x)(Vx \rightarrow Rx) \vee Dx$
2. $(y)[Vy \rightarrow (Ry \vee Dy)]$
3. $(x)[Vx \rightarrow (Rx \& Dx)]$

Choice 1, turn to frame 126.

2, turn to frame 127.

3, turn to frame 128.

126

You chose '$(x)(Vx \rightarrow Rx) \vee Dx$' as the symbolization of 'Every voter is a Republican or a Democrat'. Perhaps you should review the discussion about the placement of parentheses and the scope of quantifiers in frame 122.

Your formula is a disjunction of two parts: '$(x)(Vx \rightarrow Rx)$' and 'Dx'. Thus we would interpret this as

Every voter is a Republican *or* x is a Democrat.

This is not a sentence because of the *free* 'x'. In other words, because of the placement of the parentheses, the 'x' in 'Dx' is not *bound* to the 'x' in 'Rx' by the quantifier.

Return to frame 125 and try again.

127

'$(y)[Vy \rightarrow (Ry \vee Dy)]$' is the symbolization of 'For any individual y, if y is a voter, then y is a Republican or y is a Democrat'. The choice of individual variables is inconsequential. If 'Vx' symbolizes 'x is a voter', then 'Vy' symbolizes 'y is a voter' and 'Vz' symbolizes 'z is a voter'.

Notice that 'Every voter is a Republican or a Democrat' is in fact a false sentence. It is important to see how it is falsified: by finding one substitution instance of '$Vy \rightarrow (Ry \vee Dy)$' that is false. That is, we locate an individual who makes the antecedent true (whose name substituted for 'y' in 'Vy' yields a true sentence). We now have an individual who is a voter. We then discover that '$(Ry \vee Dy)$' is false of that individual. More explicitly, the individual voter is neither a Republican nor a Democrat. If John, a Socialist, is this individual, we have (j:John)

$$Vj \rightarrow (Rj \vee Dj).$$

This is a conditional whose *antecedent is true* but whose *consequent is false*. Hence this *conditional* is false. Since not every substitution instance is true, the universal quantification of '$Vy \rightarrow (Ry \vee Dy)$' is false.

Symbolize

Only the brave deserve the fair.

Bx: x is brave.

Dx: x deserves the fair.

____

Go to frame 129.

128

'$(x)[Vx \rightarrow (Rx \,\&\, Dx)]$' symbolizes 'For any x, if x is a voter, then x is BOTH a Republican AND a Democrat'.

We want our voter to be either one or the other.

Return to frame 125 and try again.

129

'$(x)(Dx \rightarrow Bx)$' or '$(x)(\sim Bx \rightarrow \sim Dx)$' symbolizes 'Only the brave deserve the fair'.

We must be clear on the paraphrase step. We are not claiming

> For *any* x, if x is brave, then x deserves the fair.

This means 'Every brave person deserves the fair'. We *are* claiming, on the other hand,

> For any x, x deserves the fair *only if* x is brave.

That is, either of the following two:

> For any x, if x *does* deserve the fair, then x is brave.
> For any x, if x is not brave, then x does not deserve the fair.

Here is another example:

> Some who read the book are not pleased.
> Rx: x reads the book.
> Px: x is pleased.

(There is at least one x such that x reads the book and x is *not* pleased.)

Choose the correct symbolization:

1. $(\exists x)(Rx \ \& \sim Px)$

2. $(x)(Rx \rightarrow \sim Px)$

Number 1, '$(\exists x)(Rx \ \& \sim Px)$', is correct. Number 2 is a symbolization of 'Everyone who reads the book is not pleased'.

Go to the next frame.

130

Circle the formula that correctly symbolizes the given English sentence.

All those who are wealthy or who own land are opposed to the plan if they are not stupid.

Wx: x is wealthy.
Ox: x owns land.
Px: x is opposed to the plan.
Sx: x is stupid.

1. $(x)[(Wx \vee Ox \,\&{\sim}Sx) \rightarrow Px]$
2. $(\exists x)[(Wx \vee Ox) \rightarrow ({\sim}Sx \,\&\, Px)]$
3. $(x)[(Wx \vee Ox) \rightarrow (Px \rightarrow {\sim}Sx)]$
4. $(x)[(Wx \vee Ox) \rightarrow ({\sim}Sx \rightarrow Px)]$

---

4. $(x)[(Wx \vee Ox) \rightarrow ({\sim}Sx \rightarrow Px)]$

A kiwi is a rare bird.

Kx: x is a kiwi.
Rx: x is rare.
Bx: x is a bird.

1. $(\exists x)(Kx \,\&\, Rx \,\&\, Bx)$
2. $(x)[Kx \rightarrow (Rx \,\&\, Bx)]$
3. $(x)[(Rx \,\&\, Bx) \rightarrow Kx]$
4. $(\exists x)[Kx \rightarrow (Rx \,\&\, Bx)]$

---

2. $(x)[Kx \rightarrow (Rx \,\&\, Bx)]$

Go to the next frame.

131

Circle a reasonable English interpretation of the given formula.

$(x)[(Sx \& \sim Tx) \rightarrow Ox]$

Sx: x is a society.
Tx: x is totalitarian.
Ox: x is tolerable.

1. Any totalitarian society is intolerable.

2. Every tolerable society is not totalitarian.

3. A society that is not totalitarian is tolerable.

4. No tolerable society is totalitarian.

---

3. A society that is not totalitarian is tolerable.

$\sim(\exists y)(Oy \& My \& Ly)$

Oy: y is oviparous.
My: y is a mammal.
Ly: y can fly.

1. There is no oviparous mammal that can fly.

2. It is not the case that all mammals that can fly are oviparous.

3. Some oviparous mammals cannot fly.

---

1. There is no oviparous mammal that can fly.

Go to the next page.

TEAR THIS SHEET OUT

Individual constants: $a, b, c, \ldots, m$
Individual variables: u, v, w, x, y, z
Predicates: $A, B, C, D, E, I, J, \ldots, Z$
All F's are G's: $(x)(Fx \rightarrow Gx)$
Some F's are G's: $(\exists x)(Fx \ \& \ Gx)$

-------------------------- **fold here** --------------------------

132

How are you doing?

1. '______ has a headache' is a
   (a) name
   (b) predicate
   (c) sentence
   (d) quantifier

____

2. Names of individuals are abbreviated by (capital/lower-case) letters.

3. Predicate letters followed by the appropriate number of individual variables are *abbreviations* of

____________ ____________

4. A universally quantified sentence is false if ____________

____

5. A sentence of the form 'Some M are not S' is symbolized:
   (a) $(x)(Mx \rightarrow \sim Sx)$
   (b) $(\exists x)(Mx \rightarrow \sim Sx)$
   (c) $(x)(Mx \& \sim Sx)$
   (d) $(\exists x)(Mx \& \sim Sx)$

____

---

1. predicate
2. lower-case
3. sentential functions
4. one substitution instance of the sentential function is false.
5. $(\exists x)(Mx \& \sim Sx)$

Go to the next frame.

133

Suppose $\sim(\exists x)Fx$. This can be read

(1) It is not the case that for some x, Fx.
(2) There is not even one x such that Fx.
(3) There is no x such that Fx.

Under this supposition, is $(\exists x)Fx$ true or false?

Since $(\exists x)Fx$ is *false*, by the truth conditions for an existentially quantified sentence, every substitution instance of Fx is false.

What, now, is the truth-value of *every* substitution instance of $\sim Fx$?

True / False ______

True, since if p is false, $\sim p$ is true.

If *every* substitution instance of $\sim Fx$ is true, then $(x)\sim Fx$ is true. Hence,

If $\sim(\exists x)Fx$ then ____________

If $\sim(\exists x)Fx$ then $(x)\sim Fx$.

Go to the next frame.

134

Let us assume that $(x){\sim}Fx$ is true.
(1) Every substitution instance of $\sim Fx$ is (true/false).
(2) If every instance of $\sim Fx$ is *true*, every instance of Fx is (true/false).
(3) If every instance of Fx is *false*, then there is no instance of Fx that is true.
What then is the truth-value of ${\sim}(\exists x)Fx$? (True/False)

True

You have just followed the argument to show

If $(x){\sim}Fx$ then ${\sim}(\exists x)Fx$.

We reasoned earlier that

If ${\sim}(\exists x)Fx$ then $(x){\sim}Fx$.

Thus we have discovered the *logical equivalence* between

$(x){\sim}Fx$ and ${\sim}(\exists x)Fx$.

Consider now a different sentence. What do we claim when we assert ${\sim}(x)Fx$?

1. Nothing satisfies the predicate F.

2. Not everything satisfies the predicate F.

____

If you picked choice 1, turn to frame 135.
2, turn to frame 136.

135

$\sim(x)Fx$ is the denial of $(x)Fx$.

$(x)Fx$ claims 'Every individual x is such that Fx'. To deny this is to claim '*Not every* individual x is such that Fx'. This is *not* to say '*No* individual x is such that Fx'.

For example, if I deny 'Everything is blue', I am not claiming 'Nothing is blue'. All I am claiming is that not all things in the universe are blue.

Go to the next frame.

136

$\sim(x)Fx$ claims that 'Not every individual x is such that Fx'.

But if not every individual x is such that Fx, not every substitution instance of Fx is true, so at least one instance of Fx is *false*.

And if p is false, then $\sim p$ is true, so *that* individual which does *not* satisfy F *does* satisfy $\sim F$. In other words, instances of $\sim Fx$ are true where those instances of Fx are false, and furthermore some instance of Fx *is* false.

Concretely, if not everything is blue, then some (at least one) thing is non-blue. That is,

If $\sim(x)Bx$ then $(\exists x)\sim Bx$.

There is in fact a logical equivalence between sentences of the forms:

$\sim(x)Fx$ and $(\exists x)\sim Fx$.

Our previously discovered logical equivalence was between

$(x)\sim Fx$ and

1. $(\exists x)\sim Fx$

2. $\sim(x)Fx$

3. $\sim(\exists x)Fx$

4. $\sim(\exists x)\sim Fx$

(Hint: Read these aloud in English.)

____

---

3. $\sim(\exists x)Fx$ (See frame 134.)

Go to the next frame.

137

$(\exists x)\sim Fx$ is logically equivalent to $\sim(x)Fx$
$(x)\sim Fx$ is logically equivalent to $\sim(\exists x)Fx$

A simple way to remember these equivalences is in the form of a mechanical transformation rule.

> *One can move a negation sign through a quantifier (from right to left, or left to right) by changing the quantity of the quantifier to obtain a logically equivalent sentence.*

According to this rule,

> $(x)\sim\sim Fx$ is equivalent to $\sim(\exists x)\sim Fx$
>
> and $\sim(\exists x)\sim Fx$ is equivalent to $\sim\sim(x)Fx$.
>
> Also, $(\exists x)\sim\sim Fx$ is equivalent to $\sim(x)\sim Fx$
>
> and $\sim(x)\sim Fx$ is equivalent to $\sim\sim(\exists x)Fx$.

If p is logically equivalent to q, and q is equivalent to r, is p equivalent to r?

Yes / No

Yes. Turn to frame 138.

No. Turn to frame 139.

138

p and r are logically equivalent.

If Exy: x is logically equivalent to y, then for any sentences p, q, r:

$$(Epq \ \& \ Eqr) \rightarrow Epr$$

This transitivity of logical equivalence, plus our rule for moving negation signs through quantifiers by *changing* the quantity, allows us to discover a great many logical equivalences.

$(\exists x){\sim}{\sim}Fx$ is equivalent to ${\sim}(x){\sim}Fx$
${\sim}(x){\sim}Fx$ is equivalent to ${\sim}{\sim}(\exists x)Fx$

and from Part I, the double negation of any sentence is logically equivalent to the original sentence—hence ${\sim}{\sim}(\exists x)Fx$ is logically equivalent to $(\exists x)Fx$.

Circle the formula below that is logically equivalent to ${\sim}(\exists x){\sim}Fx$.

1. $(x)Fx$

2. ${\sim}{\sim}(\exists x)Fx$

3. $(\exists x)Fx$

If you picked choice 1, turn to frame 140.

2, turn to frame 141.

3, turn to frame 142.

139

If p is logically equivalent to q, p is true (false) under precisely the same conditions as q.

And if q is logically equivalent to r, q is true (false) under precisely the same conditions as r.

Hence, p, q, and r are true (false) under precisely the same conditions.

Therefore, p and r are true (false) under precisely the same conditions, so p and r *are logically equivalent.*

Turn back to frame 138.

140

Correct. $\sim(\exists x)Fx$ is equivalent to $\sim\sim(x)Fx$, which in turn is equivalent to $(x)Fx$.

Remember that this discussion is on a level of complete generality. F has been used as a *predicate* variable ranging over any predicate no matter how complex. This degree of abstraction has prevented the actual structure of the predicate from confusing us.

Now let us look at the structure of the sentential function following the quantifiers. 'Some men are tall' is logically equivalent to 'Some tall things are men'. Symbolically, '$(\exists x)(Mx \;\&\; Tx)$' is logically equivalent to '$(\exists x)(Tx \;\&\; Mx)$'. $(\exists x)(Fx \vee Gx)$ is logically equivalent to which of these?

1. $(\exists x)(Fx \;\&\; Gx)$

2. $(\exists x)(Gx \vee Fx)$

If you picked choice 1, turn to frame 143.

2, turn to frame 144.

141

No. You moved the second negation sign in $\sim(\exists x)\sim Fx$ through the quantifier to the left, but you did not *change the quantity* of the quantifier.

Return to frame 138 and try again.

142

No. A double negation consists of two negation signs *side by side*. There is *no* double negation in $\sim(\exists x)\sim Fx$.

Hint: Move the second negation sign to the left *through the quantifier*. Then you will have a double negation.

Return to frame 138 and try again.

143

Wrong. For example, say the variable 'x' ranges over individual numbers, and $(\exists x)(Fx \vee Gx)$ is, for instance, 'Some numbers are even or odd', or '$(\exists x)(Ex \vee Ox)$'.

But '$(\exists x)(Ex \& Ox)$' would symbolize 'Some numbers are even AND odd'.

Go to the next frame.

144

$(\exists x)(Fx \vee Gx)$ is logically equivalent to $(\exists x)(Gx \vee Fx)$. And all the following pairs are logically equivalent *except one*.

1. $(x){\sim}{\sim}Fx$	$(x)Fx$
2. $(\exists x){\sim}(Fx \,\&\, Gx)$	$(\exists x)({\sim}Fx \vee {\sim}Gx)$
3. $(x)(Fx \rightarrow Gx)$	$(x)(Gx \rightarrow Fx)$
4. $(y)(Fy \leftrightarrow Gy)$	$(y)(Gy \leftrightarrow Fy)$

Which pair is not logically equivalent? ____________

---

3. $(x)(Fx \rightarrow Gx)$ is *not* logically equivalent to $(x)(Gx \rightarrow Fx)$, since $p \rightarrow q$ is not equivalent to $q \rightarrow p$.
   The others can easily be checked by truth tables.

${\sim}{\sim}p$	p	Double Negation
${\sim}(p \,\&\, q)$	$({\sim}p \vee {\sim}q)$	De Morgan's Law
$(p \leftrightarrow q)$	$(q \leftrightarrow p)$	

Go to the next frame.

145

The logical form of a sentential function is the logical form of a sentence that results from that sentential function when names (individual constants) replace the variables.

If we want to exhibit the logical form of

(1) $Ax \rightarrow (By \vee Cxy)$

we replace the variables by individual constants

(2) $Ak \rightarrow (Bf \vee Ckf)$

Now number 2 is a compound sentence consisting of the three *simple* sentences 'Ak', 'Bf', and 'Ckf'. The logical form of number 1 is

$p \rightarrow (q \vee r)$

because the logical form of a compound, nonquantified sentence is found by replacing the simple sentences by sentential variables 'p', 'q', 'r', and so forth.

This is just a tedious method for recognizing that

$(Ax \ \& \ Bz) \rightarrow {\sim}Dyy$

has the same logical form as which one of these?

1. $(Dxy \ \& \ Az) \rightarrow {\sim}Ay$

2. $(Bx \ \& \ By) \rightarrow {\sim}By$

____

Number 1 is correct. The logical form of number 2 is $(p \ \& \ q) \rightarrow {\sim}q$.

Go to the next frame.

146

Two sentential functions are logically equivalent if sentences of their logical forms are determined (by truth tables) to be logically equivalent.

Consider two sentential functions: $(Fx \rightarrow Gx)$ and $\sim(Fx \ \& \sim Gx)$. The logical forms of these sentential functions are

$$(p \rightarrow q) \quad \text{and} \quad \sim(p \ \& \sim q).$$

Truth-table analysis shows that sentences of these forms are logically equivalent. Fill in the table below.

p	q	$\sim q$	$p \ \& \sim q$	$\sim(p \ \& \sim q)$	$p \rightarrow q$
T	T	F			
T	F	T	T		
F	T				
F	F		F	T	

---

p	q	$\sim q$	$p \ \& \sim q$	$\sim(p \ \& \sim q)$	$p \rightarrow q$
T	T	F	F	T	T
T	F	T	T	F	F
F	T	F	F	T	T
F	F	T	F	T	T

Go to the next frame.

147

Since *sentences* of the forms $(p \rightarrow q)$ and $\sim(p \ \& \sim q)$ are logically equivalent, so are *sentential functions* of those logical forms—that is to say, $(Fx \rightarrow Gx)$ and $\sim(Fx \ \& \sim Gx)$.

> If in a *quantified* sentence, we substitute a logically equivalent *sentential function* for a sentential function occurring in the original sentence, we obtain a new sentence that is logically equivalent to the original (as long as the same variables are used with the same predicates).

If we substitute $\sim(Fx \ \& \sim Gx)$ for $(Fx \rightarrow Gx)$ in $(x)(Fx \rightarrow Gx)$, we obtain $(x)\sim(Fx \ \& \sim Gx)$, which is logically equivalent to $(x)(Fx \rightarrow Gx)$.

'Cows moo', symbolized '$(x)(Cx \rightarrow Mx)$', is logically equivalent to '$(x)\sim(Cx \ \& \sim Mx)$'. And '$(x)\sim(Cx \ \& \sim Mx)$' is logically equivalent to

1. $\sim(x)(Cx \ \& \sim Mx)$

2. $\sim(\exists x)(Cx \ \& \sim Mx)$ ____________

If you picked choice 1, turn to frame 148.

2, turn to frame 149.

148

You didn't *change the quantity* of the quantifier when you moved the negation sign to the left. Your choice '$\sim(x)(Cx \; \& \sim Mx)$' is 'Not everything is a cow and does not moo'. This is a true sentence (since not everything is a cow), but it is not equivalent to 'Cows moo'.

Go to the next frame.

149

'Cows moo' is equivalent to '$\sim(\exists x)(Cx \ \& \sim Mx)$', which is the sentence 'There is not a cow that does not moo'.

Often we are not interested so much in logical equivalents to a given sentence but rather in a *contradictory* sentence. What sentence *contradicts* 'Cows moo'?

There *is* a cow ______________________
(Fill in.)

There *is* a cow that does not moo.

Go to the next frame.

150

When we seek a contradictory formula to a *universally* quantified formula, we usually rely on one of these two logical equivalences:

(1) $\sim(p \rightarrow q)$ $(p \ \& \sim q)$

(2) $\sim(p \rightarrow \sim q)$ $(p \ \& \ q)$

Denying '$(x)(Cx \rightarrow Mx)$', 'Cows moo', we obtain '$\sim(x)(Cx \rightarrow Mx)$', 'Not all cows moo'. Moving the negation sign to the right through the quantifier yields '$(\exists x)\sim(Cx \rightarrow Mx)$, 'There is something such that it is not the case if it is a cow then it moos'.

Now, using number 1 above, we get '$(\exists x)(Cx \ \& \sim Mx)$', 'There is a cow who does not moo'.

Number 2 would be used in working with a sentence such as 'No pig swims', '$(x)(Px \rightarrow \sim Sx)$'. Denying this gives us '$\sim(x)(Px \rightarrow \sim Sx)$'.

What is the final sentence we obtain, using number 2 above?

1. $(\exists x)(Px \ \& \ Sx)$

2. $(x)(Px \ \& \ Sx)$

If you picked Choice 1, turn to frame 151.

2, turn to frame 152.

151

Correct.

$$\sim(x)(Px \rightarrow \sim Sx)$$
$$(\exists x) \sim (Px \rightarrow \sim Sx)$$

and by number 2

$$(\exists x)(Px \ \& \ Sx).$$

This sentence is ‘Some pigs swim’, which contradicts our original sentence ‘No pig swims’.

Turn to frame 153.

152

You neglected to change the quantity of the quantifier when you moved the negation sign to the right.

$$\sim(x)(Px \rightarrow \sim Sx)$$

becomes

$$(\exists x) \sim (Px \rightarrow \sim Sx)$$

and using number 2, we get '$(\exists x)(Px \; \& \; Sx)$', which is the sentence 'Some pigs swim'. This contradicts our original sentence, 'No pig swims'.

Go to the next frame.

153

Let's look at a diagram:

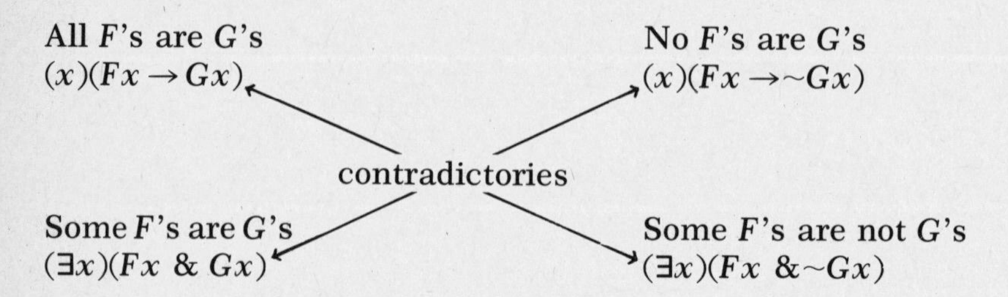

The *negation* of one corner can be shown to be logically equivalent to the diagonally opposite corner.

Before moving on to new material, let's review what we have learned about logical equivalences and about contradictories.

If we have a quantified sentence with a negation sign on either side of the quantifier, we can move the negation sign __________

____

____

and obtain a logically equivalent sentence.

We can move the negation sign *through the quantifier changing the quantity of the quantifier* and obtain a logically equivalent sentence.

Go to the next frame.

154

If two sentential functions are logically equivalent, we can ______

____

in a quantified sentence and obtain a logically equivalent sentence (if there is no change in the variables).

We can *substitute one for the other* in a quantified sentence and obtain a logically equivalent sentence.

Go to the next frame.

155

If two sentences are logically equivalent, the negation of one

________________________ the other.

---

contradicts; is contradictory to

Go to the next frame.

156

A. Write the English sentence that contradicts the sentence below.
No politician is both honest and rich.

____

Some politicians are both honest and rich.

B. Which *two* of the following symbolize the English sentence
No politician is both honest and rich.
Px: x is a politician.
Ox: x is honest.
Rx: x is rich.

1. $\sim(x)[Px \rightarrow (Ox \ \& \ Rx)]$

2. $(x)[(Px \ \& \sim Ox) \rightarrow Rx]$

3. $(x)[Px \rightarrow (\sim Ox \vee \sim Rx)]$

4. $(\exists x) \sim [Px \ \& \ (Ox \ \& \ Rx)]$

5. $\sim(\exists x)[Px \ \& \ (Ox \ \& \ Rx)]$

6. $(x)[Px \rightarrow (\sim Ox \ \& \sim Rx)]$

Note: $\sim(p \ \& \ q)$, $(\sim p \vee \sim q)$, $(p \rightarrow \sim q)$ are equivalent formulas.

____

____

3. $(x)[Px \rightarrow (\sim Ox \vee \sim Rx)]$
5. $\sim(\exists x)[Px \ \& \ (Ox \ \& \ Rx)]$

If you are incorrect, go to the next frame.

If you are correct, turn to frame 158.

157

Part A is really commonsensical. If you were to argue against the truth of 'No politician is both honest and rich' you would point to some (at least one) politician who *was* both honest and rich.

Part B is more difficult. There are several ways to get at this problem. The easiest, I think, is to *deny the denial*. That is, deny 'There is a politician who is both honest and rich'—choice 5:

$$\sim(\exists x)[Px \;\&\; (Ox \;\&\; Rx)]$$

This is logically equivalent to each of the following:

$(x) \sim [Px \;\&\; (Ox \;\&\; Rx)]$	$\sim(p \;\&\; q)$
$(x)[Px \rightarrow \sim(Ox \;\&\; Rx)]$	$(p \rightarrow \sim q)$
$(x)[Px \rightarrow (\sim Ox \vee \sim Rx)]$	$\sim(p \;\&\; q)$ goes to $(\sim p \vee \sim q)$
(choice 3)	

Another way to symbolize the given sentence is to appeal to the general form

No *F*'s are *G*'s: $(x)(Fx \rightarrow \sim Gx)$.

This gives immediately '$(x)[Px \rightarrow \sim(Ox \;\&\; Rx)]$'. This sentence can be manipulated in several ways to yield numbers 3 and 5.

Go to the next frame.

158

Circle the symbolizations of the English sentences:

> Jx: x is a junior.
> Sx: x submitted a paper.
> Px: x passed the course.

Every junior who submitted a paper passed the course.

1. $(x)(Jx \& Sx \& Px)$
2. $(x)[Jx \leftrightarrow (Sx \& Px)]$
3. $(x)[Jx \rightarrow (Sx \& Px)]$
4. $(x)[(Jx \& Sx) \rightarrow Px]$

---

4. $(x)[(Jx \& Sx) \rightarrow Px]$ That is, 'For every x, *if* x is a junior *and* x submitted a paper *then* x passed the course'.

No junior who submitted a paper passed the course.

1. $\sim(x)[(Jx \& Sx) \rightarrow Px]$
2. $(x)[Jx \rightarrow \sim(Sx \& Px)]$
3. $(x)[(Jx \& Sx) \rightarrow \sim Px]$
4. $\sim(x)[Jx \rightarrow (Sx \& Px)]$

---

3. $(x)[(Jx \& Sx) \rightarrow \sim Px]$ That is, 'For any x, *if* x is a junior who submitted a paper, *then* x did not pass the course'. (OR 'Every junior who submitted a paper did not pass the course'.)

Go to the next frame.

159

Consider '$(x)[(Jx \& Sx) \rightarrow Px]$' and '$(x)[(Jx \& Sx) \rightarrow \sim Px]$'.
Circle which of the following applies:

1. logically equivalent

2. contradictory

3. neither

If you picked choice 1, turn to frame 160.

2, turn to frame 161.

3, turn to frame 162.

160

Wrong. '$(x)[(Jx \& Sx) \rightarrow Px]$' and '$(x)[(Jx \& Sx) \rightarrow \sim Px]$' can't be true (false) under precisely the same conditions. Let us consider a situation where these two do not have the same truth-value.

Suppose we found an individual, say d, who satisfied '$(Jx \& Sx)$'. That is, suppose d was a junior who submitted a paper. Now, either Pd or $\sim Pd$—that is to say, either d passed the course or he didn't.

In one case, '$(Jd \& Sd) \rightarrow Pd$' is true, and '$(Jd \& Sd) \rightarrow \sim Pd$' false, and in the other case, the truth-values of these sentences—substitution instances of '$(Jx \& Sx) \rightarrow Px$' and '$(Jx \& Sx) \rightarrow \sim Px$'—are reversed.

Return to frame 159 and choose another answer.

161

Not quite. If '$(x)[(Jx \& Sx) \rightarrow Px]$' is false, it does not follow that '$(x)[(Jx \& Sx) \rightarrow \sim Px]$' is true.

In English, if 'All juniors who submitted a paper passed the course' is false, it does not mean that no juniors who submitted a paper passed the course. After all, some of those juniors may have passed, and some may not have passed.

Return to frame 159 and choose the correct answer.

162

Correct. They are clearly not logically equivalent. They are not contradictory, since the contradictory of

$$(x)[(Jx \ \& \ Sx) \rightarrow Px],$$

which is the sentence 'Every junior who submitted a paper passed the course', is the sentence

$$\sim(x)[(Jx \ \& \ Sx) \rightarrow Px],$$

which is the sentence 'Not every junior who submitted a paper passed the course'.

Our original sentences are of the forms

All *A* are *B* and No *A* are *B*

and these are neither equivalent nor contradictory.

After all this abstract material you might want to take a rest. Here is a good place to stop and mull over the concepts of logical equivalence and contradiction.

When you return to this program you might wish to review the material in this section. If so, begin with frame 133 and run through this part of the program again. It should not take as long the second time through.

The program continues in the next frame.

163

Would you like to test yourself to see if you have a firm grasp on the material in the previous section? Circle the correct choice.

1. 'No G is H', symbolized '$(x)(Gx \to {\sim}Hx)$', is contradictory to

(a) All G is H $(x)(Gx \to Hx)$

(b) Some G is H $(\exists x)(Gx \;\&\; Hx)$

(c) Some G is not H $(\exists x)(Gx \;\&\; {\sim}Hx)$

2. ${\sim}(\exists x){\sim}{\sim}Fx$ is logically equivalent to

(a) $(x){\sim}Fx$

(b) $(\exists x){\sim}Fx$

(c) ${\sim}(x)Fx$

3. 'If every senior at the party was not drunk, then some freshmen were happy' should be symbolized

Sx: x is a senior
Px: x was at the party
Dx: x was drunk
Ry: y is a freshman
Ay: y is happy

(a) $(x)(y)[(Sx \;\&\; Px \;\&\; {\sim}Dx \;\&\; Ry) \to Ay]$

(b) $(x)(Sx \;\&\; Px \;\&\; {\sim}Dx) \to (y)(Ry \to Ay)$

(c) $(x)(Sx \;\&\; Px \;\&\; {\sim}Dx) \to (\exists y)(Ry \to Ay)$

(d) $(x)[(Sx \;\&\; Px) \to {\sim}Dx] \to (\exists y)(Ry \;\&\; Ay)$

4. Suppose you know that two *sentential functions* Fx and Gx are logically equivalent. What is the relation between

$${\sim}(x)Fx \quad \text{and} \quad (\exists x){\sim}Gx?$$

(a) logically equivalent

(b) contradictory

(c) neither

---

1. (b)
2. (a)
3. (d)
4. (a)

Go to the next frame.

164

When faced with the problem of determining if two symbolic formulas are logically equivalent, how would you begin?

A. I would begin by moving negation signs through the quantifiers of one of the formulas.

B. I would begin by replacing the sentential functions following the quantifiers by some logically equivalent sentential functions.

C. I do not want to commit myself ahead of time. I would begin as in A or B depending on the formulas in question.

Some people adopt a standard procedure for all problems, and always try one attack first. Choices A and B are not incorrect, but the flexibility in approach C has much to be said for it. There is no right way to begin, since there is unfortunately no *general* procedure for determining when two formulas are logically equivalent.

Go to the next frame.

165

There are many more techniques for proving that two formulas are logically equivalent besides playing around with negation signs or substituting sentential functions whose equivalence can be determined by truth-functional analysis. You should, if you are interested, consult some other text for these proof procedures.

The following fact may be of passing interest to you:

> If two formulas are logically equivalent, then it is possible to construct a proof showing that they are.

(The statement just made is itself provable.)

Do you suppose that there is a *general* method for showing that two formulas are contradictory?

Yes/No

If your answer is yes, turn to frame 166.

no, turn to frame 167.

166

No, there is not.

If there was a general method for determining contradictories, there would then be a general method for determining logical equivalences.

Do you see why this would be so? Think about it a moment and then turn to the next frame for a fuller explanation.

167

There cannot be a general procedure for determining contradictories (GPC).

I have already stated (but not proved) that there is no general procedure for determining logical equivalence (GPLE). And if there were a GPC then there would be a GPLE.

If p and q are logically equivalent, then p and $\sim q$ are contradictory, and conversely. Our GPLE would be as follows:

To determine if p and q are logically equivalent, use GPC on p and $\sim q$ to determine if they are contradictory. If they are, then p and q are logically equivalent.

Show the following sentences are contradictory:

1. $(\exists x)(Ax \vee Bx \vee Cx)$

2. $(x){\sim}[{\sim}Ax \rightarrow (Bx \vee Cx)]$

There are several correct ways of doing this, of course. I found the simplest procedure to be to show that number 1 is logically equivalent to the denial of number 2. The denial of number 2 is '$\sim(x){\sim}[{\sim}Ax \rightarrow (Bx \vee Cx)]$'.

What was my next step?

Go to the next frame.

168

1. $(\exists x)(Ax \vee Bx \vee Cx)$
Denial of 2. $\sim(x)\sim[\sim Ax \rightarrow (Bx \vee Cx)]$

My next step is either to replace '$\sim(x)\sim$' by '$(\exists x)$', or to move the second negation sign one place to the left,

2a. $\sim\sim(\exists x)[\sim Ax \rightarrow (Bx \vee Cx)]$

and then by double negation:

2b. $(\exists x)[\sim Ax \rightarrow (Bx \vee Cx)]$

Now all I have to do is show the sentential functions following '$(\exists x)$' in numbers 1 and 2b are equivalent. We can insert parentheses in number 1 any way we please, and so we find that we must show the logical equivalence between

$[Ax \vee (Bx \vee Cx)]$

and

$[\sim Ax \rightarrow (Bx \vee Cx)]$.

I could set up truth tables to show the logical equivalence between

$[p \vee (q \vee r)]$ and $[\sim p \rightarrow (q \vee r)]$.

But do you see a further simplification of the problem?

Go to the next frame.

169

Since '$(Bx \vee Cx)$' occurs in both sentential functions as a unit, I could determine the equivalence between $(p \vee q)$ and $(\sim p \rightarrow q)$.

Since 'p' and 'q' are variables for any sentence (sentential function), I could let p be 'Ax' and q be '$(Bx \vee Cx)$'.

In any event, we show by truth-table analysis the equivalence between $[p \vee (q \vee r)]$ and $[\sim p \rightarrow (q \vee r)]$ or $(p \vee q)$ and $(\sim p \rightarrow q)$.

Do it!

This tells us that '$[Ax \vee (Bx \vee Cx)]$' and '$[\sim Ax \rightarrow (Bx \vee Cx)]$' are logically equivalent. And thus so are '$(\exists x)[Ax \vee (Bx \vee Cx)]$' and '$(\exists x)[\sim Ax \rightarrow (Bx \vee Cx)]$'. But the last-mentioned sentence is equivalent to '$\sim(x)\sim[\sim Ax \rightarrow (Bx \vee Cx)]$', which is the *negation* of our original sentence '$(x)\sim[\sim Ax \rightarrow (Bx \vee Cx)]$'.

Since one sentence is equivalent to the negation of the other, the original sentences are contradictory.

Go to the next frame.

170

Are the following two sentences logically equivalent?

1. $(x)(Ax \rightarrow Bx) \rightarrow (\exists y)Cy$

2. $(\exists y)Cy \lor (\exists x)(Ax \,\&\, {\sim}Bx)$

Yes / No

If your answer is yes, turn to frame 171.

no, turn to frame 172.

171

Correct. If you just *guessed,* go on to the next frame, where it is worked out in detail.

If you really worked out this difficult problem, you have my congratulations, and you may turn to frame 173.

172

1. $(x)(Ax \rightarrow Bx) \rightarrow (\exists y)Cy$
2. $(\exists y)Cy \vee (\exists x)(Ax \,\&{\sim}Bx)$

These are logically equivalent.
First of all, $(p \vee q)$ is equivalent to $(q \vee p)$, so switch number 2 to

2a. $(\exists x)(Ax \,\&{\sim}Bx) \vee (\exists y)Cy$.

Now you just worked out a truth table showing that $({\sim}p \rightarrow q)$ and $(p \vee q)$ are equivalent, so number 2a is equivalent to

2b. ${\sim}(\exists x)(Ax \,\&{\sim}Bx) \rightarrow (\exists y)Cy$.

All you have to do now is show the *antecedents* of numbers 1 and 2b to be equivalent. That is, show

$(x)(Ax \rightarrow Bx)$ and ${\sim}(\exists x)(Ax \,\&{\sim}Bx)$

to be equivalent. Move the negation sign to the right, obtaining

$(x){\sim}(Ax \,\&{\sim}Bx)$.

Now show that $(p \rightarrow q)$ is equivalent to ${\sim}(p \,\&{\sim}q)$ by truth tables, and you are finished.

Go to the next frame.

173

Match the symbolization with its appropriate English sentence. (Take the universe restricted to people only.)

1. $(x)Ljx$ ____________ a. John likes someone.
2. $(\exists x)Lxj$ ____________ b. Everyone likes John.
3. $(x)Lxj$ ____________ c. John likes everyone.
4. $(\exists x)Ljx$ ____________ d. Someone likes John.

---

1. $(x)Ljx$ c. John likes everyone.
2. $(\exists x)Lxj$ d. Someone likes John.
3. $(x)Lxj$ b. Everyone likes John.
4. $(\exists x)Ljx$ a. John likes someone.

Go to the next frame.

174

1. '$(\exists x)Ljx$' symbolizes 'John likes someone'.
   '$(\exists y)(\exists x)Lyx$' symbolizes 'Someone likes someone'.

2. '$(x)Ljx$' symbolizes 'John likes everyone'.

   '$(y)(x)Lyx$' symbolizes ______________________________
   (Fill in the blank.)

'$(y)(x)Lyx$' symbolizes 'Everyone likes everyone'.

Go to the next frame.

175

1. '$(\exists x)(y)Lxy$' can be expanded partially to read
   (a) There is *some x* such that for *any y*, that x likes y. THAT IS:
   (b) Someone likes everyone.

2. '$(\exists y)(x)Lxy$', when the quantifiers are stated in English, becomes
   (a) There is *some y* such that for *any x*, x likes that y. THAT IS:
   (b) There is someone whom everyone likes.

3. '$(x)(\exists y)Lxy$' is to be read

   (a) ______________________ such that x likes that y.
   Fill in like lines (a) in the above examples.

For *any x*, there is *some y*, such that x likes that y.

Circle the English sentence below which is symbolized in example number 3.

Someone *is liked by* everyone.

Everyone *likes* someone (or other).

Everyone *likes* someone (or other).

Notice the difference between symbolizations 2 and 3, and notice how these sentences differ in meaning.

Go to the next frame.

176

1. John likes everyone who studies philosophy.
   (a) For any x, if x studies philosophy, then John likes x.
   (b) $(x)(Sx \rightarrow Ljx)$

2. John likes only those who study philosophy.
   (a) For any x, John likes x *only if* x studies philosophy.

Circle the correct symbolization of number 2a.

$(x)(Ljx \rightarrow Sx)$

$(x)(Sx \rightarrow Ljx)$

$(x)(\sim Sx \rightarrow Ljx)$

'$(x)(Ljx \rightarrow Sx)$' is correct.
'$(x)(Sx \rightarrow Ljx)$' is the symbolization of number 1.

Go to the next frame.

177

'$(x)(Sx \rightarrow Ljx)$' symbolizes 'John likes anyone who studies philosophy'.

What does '$(\exists y)(x)(Sx \rightarrow Lyx)$' symbolize?

1. John likes someone who studies philosophy.

2. Everyone likes someone who studies philosophy.

3. Someone likes anyone who studies philosophy.

---

3. Someone likes anyone who studies philosophy.

Go to the next frame.

178

1. John likes no one who studies philosophy.
   (a) For anyone who studies philosophy, John does not like him.
   (b) For any x, if Sx, then $\sim Ljx$.
   (c) $(x)(Sx \rightarrow \sim Ljx)$

2. John likes no one who hits his (John's) sister.
   Sxy: x is a sister of y. (Remember: Syx: y is a sister of x.)
   Ixy: x hits y.

   (a) For any x, if *x hits a sister of John*, then John does not like x.
      (i) x hits a sister of John.
      (ii) There is a y, x hits y, and y is a sister of John.
   (b) For any x, if $(\exists y)(Ixy \;\&\; Syj)$, then John does not like x.
   (c) $(x)[(\exists y)(Ixy \;\&\; Syj) \rightarrow \sim Ljx]$

3. No one likes anyone who hits his sister. (That is, everyone is similar to John in number 2 above.)
   (a) For every x and every y, if y hits a sister of x, x does not like y.
      (i) y hits a sister of x.
      (ii) There is a z, y hits z, and z is a sister of x.

   (b) For every x and y, if ______________________________

   then ______________________________

   (c) $(x)\,(y)$ [______________________________]

(b) For every x and y, if $(\exists z)(Iyz \;\&\; Szx)$, then x does not like y.
(c) $(x)(y)[(\exists z)(Iyz \;\&\; Szx) \rightarrow \sim Lxy]$

Go to the next frame.

179

You must watch your parentheses and brackets in problems of this sort. When we have a sentence of the form

For all x, if there is a y, Fy, then Gx

the existential quantifier of 'y' is in the antecedent of the (conditional) sentential function:

if there is a y, Fy, then Gx.

Let us see what the difference is between

$(\exists y)Fy \rightarrow p$

and

$(\exists y)(Fy \rightarrow p)$.

Take 'If someone squeals, the cause is lost'.
(a) If $(\exists y)Sy$, then C
(b) $(\exists y)Sy \rightarrow C$

But suppose we write '$(\exists y)(Sy \rightarrow C)$'. This is a true sentence if there is one name which substituted for 'y' in '$(Sy \rightarrow C)$' yields a true sentence. Let us consider Fido, f, who is *not* a squealer. Then '$(Sf \rightarrow C)$' is a true sentence because the antecedent 'Sf' is false. The name of any non-squealer in the universe put in place of 'y' gives a true sentence.

In other words, '$(\exists y)(Fy \rightarrow p)$' is true if '$(Fy \rightarrow p)$' is true for *some* choice of 'y'. And any choice of 'y' for which 'Fy' is false will make '$(Fy \rightarrow p)$' true.

On the other hand, '$(\exists y)Fy \rightarrow p$' is a false sentence in those cases (and only those cases) where '$(\exists y)Fy$' is a true sentence and 'p' is a false one. That Fido is a non-squealer does not make '$(\exists y)Sy$' ('Someone is a squealer') false.

Go to the next frame.

180

No one likes anyone who hits his own sister.
Lxy: x likes y.

(a) No x likes any y if y hits his own sister.

(i) y hits his own sister.
(ii) $(\exists z)(Iyz \;\&\; Szy)$

(b) For any x and any y, if ____________ then ____________

(c) ____
(Symbolize completely.)

(b) For any x and any y, if $(\exists z)(Izy \;\&\; Szy)$ then $\sim Lxy$.
(c) $(x)(y)[(\exists z)(Iyz \;\&\; Szy) \rightarrow \sim Lxy]$

Go to the next frame.

181

In symbolizing, we paraphrase, making subject terms explicit. We then symbolize the shorter complex clauses inside the sentence and then work out to the whole sentence.

There is a book that is read by every student who reads any book at all.

Bx: x is a book.
Sx: x is a student.
Rxy: x reads y.

First we paraphrase the sentence:

There is some x such that (x is a book *and* every y who is a student and who has read any book at all reads x).

Then we symbolize 'y has read any book at all' (that is to say, 'y has read some book or other'):

$(\exists z)(Ryz \,\&\, Bz)$

or equally correct

$(\exists z)(Bz \,\&\, Ryz)$.

Here is a paraphrase with partial symbolization:

$(\exists x)$(x is a book & every y who is a student and $(\exists z)(Ryz \,\&\, Bz)$ reads x).

Now symbolize 'Every y who is a student reads x'.

$(y)(Sy$ ____________________

$(y)(Sy \rightarrow Ryx)$

Go to the next frame.

182

Given that 'Every y who is a student reads x' is symbolized

$$(y)(Sy \rightarrow Ryx)$$

the symbolization of

Every y who is a student and $(\exists z)(Ryz \; \& \; Bz)$ reads x

is

$$(y)\{[Sy \; \& \; (\exists z)(Ryz \; \& \; Bz)] \rightarrow Ryx\}.$$

Now symbolize

There is an x such that [(x is a book) *and* (every student who reads any book at all reads x)].

$(\exists x)$ ____________________

$(\exists x) \big(Bx \; \& \; (y)\{[Sy \; \& \; (\exists z)(Ryz \; \& \; Bz)] \rightarrow Ryx\}\big)$

Go to the next frame.

183

We have seen that the standard symbolic form for such sentences as

All F's are G's

is

$(x)(Fx \rightarrow Gx)$.

1. All seniors are eligible.

$(x)(Sx \rightarrow Ex)$

If we restrict those F's that are G's by adding a further condition H, (All those F's that are H's are G's), we obtain the standard form

$(x)[(Fx \ \& \ Hx) \rightarrow Gx]$.

2. Those seniors who have had chemistry are eligible.

$(x)[(Sx \ \& \ Cx) \rightarrow Ex]$

This sample sentence is clearly a different assertion from

3. All those who are eligible are seniors who have had chemistry.

$(x)[Ex \rightarrow (Sx \ \& \ Cx)]$

This can also be stated in English as

Only seniors who have had chemistry are eligible.

A universally quantified sentence does not assert (although it may presuppose) the existence of anything. In the above examples, every one of the sentences could as a matter of fact be true, and yet in fact there may not *exist* one eligible person.

Existentially quantified sentences, on the other hand, do assert the existence of at least one thing satisfying one condition, or *conjointly* more than one condition.

1. Some seniors are eligible.

$(\exists x)(Sx \ \& \ Ex)$

2. Some seniors who have had chemistry are eligible.

$(\exists x)[(Sx \ \& \ Cx) \ \& \ Ex]$

This makes the same existence claim as

3. Some seniors who are eligible have had chemistry.

$(\exists x)[(Sx \ \& \ Ex) \ \& \ Cx]$

Go to the next frame.

184

1. Gold Dollar is the favorite horse.

'Gold Dollar' is a name of some individual horse and so will be abbreviated by an individual constant—say, 'g'. We might be tempted to say that sentence 1 asserts that g has the property of being the favorite horse, and thus symbolize our sentence 'Ag'.

But look at

2. The favorite horse is the crowd's choice.

Here 'the favorite horse' is a denoting expression referring to an individual. Shall we abbreviate this expression as 'f' and sentence 2 as 'Cf'? This will hardly do, since from sentences 1 and 2 we *should* be able to infer

3. Gold Dollar is the crowd's choice.

And we see no relation between 'Ag' and 'Cf' that permits this (symbolized 'Cg').

Sentence 1 really asserts that one individual, Gold Dollar, is

________________________ the individual the favorite horse.

(Fill in the relationship.)

---

identical to (or the same as)

Go to the next frame.

185

Logicians use the familiar equals sign to symbolize the relation of *identity*.

Thus,

1. Gold Dollar is the favorite horse
   g f

2. The favorite horse is the crowd's choice.
   f c

3. Gold Dollar is the crowd's choice.

will be symbolized

1. $g = f$

2. $f = c$

3. ____________
   (Fill in.)

---

3. $g = c$

Go to the next frame.

186

Using the symbol for identity, we can now symbolize such sentences as

1. Whoever is *the* winner receives ten dollars.
   i: the winner
   Rx: x receives ten dollars.

   (a) For any x, if x is the winner, x receives ten dollars.
   (b) $(x)[(x = i) \rightarrow Rx]$

2. No more than one freshman is eligible.
   (a) If there are two freshmen who are eligible, they are the same individual.
   (b) $(x)(y)[(Fx \; \& \; Fy \; \& \; Ex \; \& \; Ey) \rightarrow (x = y)]$

3. Every member of the class *except Mary* passed.
   Cx: x is a member of the class.
   Px: x passed.
   m: Mary

   (a) For any x, if x is a member of the class and x is not Mary, then x passed.

   (b) $(x)\{[(Cx \; \&$ ________________
   (Complete the symbolization.)

$(x)\{[(Cx \; \& {\sim}(x = m)] \rightarrow Px\}$ or, of course, $(x)\{[(Cx \; \& {\sim}(m = x)] \rightarrow Px\}$

Go to the next frame.

187

Let's look at some "real life" cases. First, an example from elementary geometry.

(1) If any two lines are perpendicular to a third line, then they are parallel to each other.
(2) If any line is perpendicular to another line, then that other line is perpendicular to the first.

Thus

(3) If any two lines are not parallel, then it is not the case that there is some third line which is perpendicular to both.

We could begin rephrasing statement 1 as

For any x and y, *if x and y are lines* and . . .

but notice that only *lines* are mentioned in the three sentences above. So we can restrict our universe (of discourse) to lines. That is, the range of the individual variables will be restricted to (coplanar) lines.

1. $(x)(y)(z)[(Exz \,\&\, Eyz) \rightarrow (Axy \,\&\, Ayx)]$

2. $(x)(y)(Exy \rightarrow Eyx)$

3. For any x and y, if x is not parallel to y, then it is not the case there exists a z such that z is perpendicular to x and z is perpendicular to y.

Circle the correct symbolization below.

(a) $(x)(y) \sim [Axy \rightarrow \sim (\exists z)Ezx \,\&\, Ezy]$

(b) $(x)(y)[\sim Axy \rightarrow \sim (\exists z)(Ezx \,\&\, Ezy)]$

(c) $(x)(y)[\sim Axy \rightarrow \sim (\exists z)(Exz \,\&\, Eyz)]$

If you picked choice a, go to frame 188.

b, go to frame 189.

c, go to frame 190.

188

Incorrect. You are not watching parentheses and scope. You chose '$(x)(y) \sim [(Axy \rightarrow \sim(\exists z)Ezx \ \& \ Ezy]$'. The first negation sign here represents a denial of the whole sentential function '$[Axy \rightarrow \sim(\exists z)Ezx \ \& \ Ezy]$'. Also the scope of '$(\exists z)$' extends only to '$Ezx$' and does not include '$Ezy$'.

Schematically

$$(x)(y) \sim [Axy \rightarrow \sim(\exists z)Ezx \ \& \ Ezy].$$

Return to frame 187 and select a better answer.

189

'$(x)(y)[\sim Axy \rightarrow \sim(\exists z)(Ezx \ \& \ Ezy)]$' symbolizes 'For any x and any y, if (x is not parallel to y), then it is not the case [there exists a z such that (z is perpendicular to x and z is perpendicular to y)].

An example from philosophy is

(1) A substance is, by definition, not limited by anything.
(2) Any substance can be limited only by a different substance.
(3) If any two substances are different from one another, then one limits the other and is limited by the other.

Consequently,

(4) If any substance exists, then no substance is different from it.

Now if we treat a definition as a universally quantified sentence, we can paraphrase as follows:

1. For any x, x is a substance if and only if x is not limited by anything.

2. For any x and y, if x is a substance and y limits x then y is a substance and different from x.

3. For any x and y, if x and y are substances and x is different from y, then

____

(Complete the paraphrase of 3.)

---

x limits y and x is limited by y. (or, x limits y and is limited by y.)

---

4. For any x, if x is a substance, then ______________

____

____

(Complete the paraphrase of 4.)

---

there is no y such that y is a substance and is different from x. (or equivalently, every y which is a substance is identical to x.)

Turn to frame 191.

190

Incorrect. The formula '$(\exists z)(Exz \ \& \ Eyz)$' is interpreted as

> There is a z such that x is perpendicular to z and y is perpendicular to z.

It is true, according to statement 2, that if z is perpendicular to x then x is perpendicular to z. But what sentence 3 *actually asserts is*

> . . . it is not the case there is a z such that z is perpendicular to x and z is perpendicular to y.

So we must, consistent with our earlier symbolizing, place the 'z' first and 'x' and 'y' last.

The correct symbolization is '$(x)(y)[\sim Axy \rightarrow \sim(\exists z)(Ezx \ \& \ Ezy)]$'.

Turn back to frame 189.

191

(1) For any x, x is a substance if and only if x is *not* limited by anything.
(2) For any x and y, if x is a substance and y limits x, then y is a substance and different from x.
(3) For any x and y, if x and y are substances and x is different from y, then x limits y and y limits x.
(4) For any x, if x is a substance, then no substance y is different from x.

The symbolizations of these sentences are

1. $(x)[Sx \leftrightarrow (y)(\sim Lyx)]$

2. $(x)(y)\{(Sx \ \& \ Lyx) \rightarrow [Sy \ \& \sim(y = x)]\}$

3. ____________________________

4. $(x)\{Sx \rightarrow \sim(\exists y)[Sy \ \& \sim(y = x)]\}$ or $(x)\{Sx \rightarrow (y)[Sy \rightarrow (y = x)]\}$

Write in the correct symbolization of sentence 3 above.

(a) $(x)(y)\{[Sx \ \& \ Sy \ \& \sim(x = y)] \rightarrow (Lxy \ \& \ Lyx)\}$

(b) $(x)(y)\{[Sxy \ \& \sim(x = y)] \rightarrow (Lxy \ \& \ Lyx)\}$

If you picked choice a, go to frame 192.

b, go to frame 193.

192

Correct. The symbolization can be more easily determined if sentence 3 is reworded as

> For any x and y, if x is a substance and y is a substance and x is different from y, then x limits y and y limits x.
>
> $(x)(y)\{[Sx \ \& \ Sy \ \& \sim(x = y)] \rightarrow (Lxy \ \& \ Lyx)\}$

Go to tear-out sheet after frame 193.

193

You are mistaken.

. . . *x* and *y* are substances . . .

is a sentential function with a *multiple* subject. It should be paraphrased further to read

. . . *x* is a substance and *y* is a substance . . .

The *simple* predicate is

____ is a substance

symbolized '*S*____'.

Go back to frame 191 and see how this simple predicate is used in the symbolization of sentences 1, 2, and 4. Then choose the correct answer.

TEAR THIS SHEET OUT

The next case comes from a text in theoretical biology. Since it is rather long, you should refer to this information for the next few pages.

These are the statements we shall be symbolizing:

(1) Distinct slices of an organized unity have no parts in common.
(2) Of any two distinct slices of an organized unity, one temporally precedes the other.
(3) Between any two distinct slices of an organized unity, there is a third slice distinct from both. (Note: 'Between' means temporally between.)

Ox: x is an organized unity.

Pxy: x is a part of y.

Sxy: x is a slice of y.

Txy: x temporally precedes y.

------------------------- fold here -------------------------

194

Paraphrasing the sentences on the tear-out sheet, we obtain

(1) For any x, y, and z, if z is an organized unity and x and y are distinct slices of z, then no part of x is a part of y.

Is the following a good paraphrase of sentence 2?

For any x and y, if there is an organized unity z of which x and y are slices and x and y are distinct, then x temporally precedes y.

Yes / No

If your answer is yes, turn to frame 195.

no, turn to frame 196.

195

Statement 2 *cannot* be paraphrased as

> For any x and y, if there is an organized unity z of which x and y are slices and x and y are distinct, then x temporally precedes y.

The antecedent of the conditional is correct, but the consequent is not. We cannot say that x precedes y. What is asserted by our original sentence is that of *any* two distinct slices one of them precedes the other.

Therefore, the correct consequent reads

> . . . then either x precedes y or y precedes x.

Turn to frame 197.

196

The given paraphrase does *not* do justice to the sentence.

When we say of two things x and y that *one precedes the other*, we must amplify the phrase to 'either x precedes y or y precedes x'.

You are doing well. Turn to the next frame.

197

(3) For any x, y, and z, if z is an organized unity and x and y are distinct slices of z, then there is a w such that w is a slice of z and w is (temporally) between x and y.

From sentence 2 we know that if x and y are distinct slices of z, then one precedes the other. We shall not be making a material assumption, then, if we add to our *antecedent* the condition that it is x that precedes y.

We now have

(3) $(x)(y)(z)\{[(Oz \mathbin{\&} Sxz \mathbin{\&} Syz) \mathbin{\&} {\sim}(x = y) \mathbin{\&} Txy] \rightarrow$
$(\exists w)[Swz \mathbin{\&} {\sim}(w = x) \mathbin{\&} {\sim}(w = y) \mathbin{\&} w$ is between x and $y]\}$

Using 'T', symbolize 'w is between x and y'—illustrated below:

____

$(Txw \mathbin{\&} Twy)$ (We have already said that Txy.)

Go to the next frame.

198

Here are the paraphrases with some symbolizations:

(2) For any x and y, if there is an organized unity z such that x and y are distinct slices of z, then either x precedes y or y precedes x.

$$(x)(y)\{[(\exists z)(Oz \ \& \ Sxz \ \& \ Syz) \ \& {\sim}(x = y)] \rightarrow (Txy \vee Tyx)\}$$

(3) For any x, y, and z, if z is an organized unity and x and y are distinct slices of z and x precedes y, then there is a w such that w is a slice of z and w is between x and y.

$$(x)(y)(z)\{[(Oz \ \& \ Sxz \ \& \ Syz) \ \& {\sim}(x = y) \ \& \ Txy] \rightarrow (\exists w)[Swz \ \& {\sim}(w = x) \ \& {\sim}(w = y) \ \& \ (Txw \ \& \ Twy)]\}$$

Circle the proper symbolization of sentence 1 below:

(1) For any x, y, and z, if z is an organized unity such that x and y are distinct slices of z, then no part of x is a part of y.

A. $(x)(y)(z)\{[Oz \ \& \ Sxz \ \& \ Syz \ \& {\sim}(x = y)] \rightarrow {\sim}(w)(Pwx \rightarrow Pwy)\}$

B. $(x)(y)(z)\{[Oz \ \& \ Sxz \ \& \ Syz \ \& {\sim}(x = y)] \rightarrow (w)(Pwx \rightarrow {\sim}Pwy)\}$

If you picked choice A, turn to frame 199.

B, turn to frame 200.

199

Wrong. You have been misled by '$\sim(w)(Pwx \rightarrow Pwy)$', which should be interpreted as 'IT IS NOT THE CASE THAT EVERY w that is a part of x is also a part of y' or 'Not every part of x is a part of y'. This means that some parts of x *may* be parts of y.

We want to symbolize

No part of x is a part of y

or in other words

Every part of x is *not* a part of y.

This is symbolized '$(w)(Pwx \rightarrow \sim Pwy)$'.

Go to the next frame.

200

The correct symbolization is choice B.

An analysis of Newtonian particle mechanics provides us with a problem in interpretation.

Px: x is a particle.
Mxy: x is the mass of y.
$Vxyzw$: x is the vector force from y to z at time w.
0: the number zero

1. $(u)(x)(y)(z)(w)[(Px \;\&\; Py \;\&\; Vuxyz \;\&\; Vwyxz) \rightarrow (u = -w)]$ (Here '—' is used in the mathematical sense of minus.)

   (a) For any u, x, y, z, w, if x and y are particles and u is the vector force from x to y at time z, and w is the vector force from y to x at time z, then u equals minus w.

   (b) Between any two particles at any instant of time, there are equal and oppositely directed forces.

2. $(x)(y)[(Px \;\&\; Myx) \rightarrow (y > 0)]$ (Here '>' is used in the normal mathematical sense of 'is greater than'.)

   (a) For any x and y, if x is a particle and y is the mass of x then y is greater than zero.

   (b) ______________________________
   (Write in a reasonable English interpretation.)

The mass of any particle is greater than zero. (or, All particles have positive mass.)

Go to the next frame.

201

Another problem in interpretation can be found in the theory of preference or choice.

Pxy: The subject prefers x to y.

We have the following sentences:

1. $(x)(\exists y)Pyx$

2. $(x)(y)(Pxy \rightarrow \sim Pyx)$

3. $(x)(y)(z)[(Pxy \ \& \ Pyz) \rightarrow Pxz]$

The interpretation of sentence 1 is

For any x, ______________________________

____

(Complete without using *abbreviations*.)

For any x, there is a y such that the subject prefers y to x.

For sentence 2, the interpretation is

For any x and y, if the subject prefers x to y, then he does not prefer y to x.

And the interpretation of sentence 3 is

____

____

____

(Again use no abbreviations.)

For any x, y, z, if the subject prefers x to y and y to z, then he prefers x to z.

Go to the next frame.

202

(1) For any x, there is a y such that the subject prefers y to x.
(2) For any x and y, if the subject prefers x to y, then he does not prefer y to x.
(3) For any x, y, z, if the subject prefers x to y and y to z, then he prefers x to z.

Can we phrase these three into passable English?

1. Nothing is preferred above all else.

2. If one thing is preferred to another, then ______________________

____

If one thing is preferred to another, then *the other is not preferred to* the first.

---

3. If one thing is preferred to a second and the second ____________

____

____

____

If one thing is preferred to a second and the second *preferred to a third, then the first is preferred to the third.*

Go to the next frame.

203

Circle the correct symbolization of

Every senior knows at least one freshman.
Sx: x is a senior.
Kxy: x knows y.
Rx: x is a freshman.

1. $(x)[(Sx \ \& \ Kxy) \rightarrow (\exists y)Ry]$

2. $(x)(\exists y)[(Sx \ \& \ Kxy) \rightarrow Ry]$

3. $(x)[Sx \rightarrow (\exists y)(Kxy \ \& \ Ry)]$

4. $(x)[Sx \rightarrow (Kxy \ \& \ (\exists y)Ry)]$

---

3. $(x)[Sx \rightarrow (\exists y)(Kxy \ \& \ Ry)]$

If you are incorrect, go to frame 204.

If you are correct, go to frame 205.

204

'Every senior knows at least one freshman' is paraphrased as

> For any x *if* x is a senior *then* there is some (at least one) y such that (x knows y *and* y is a freshman).

From this, it is an easy step to

$$(x)[Sx \rightarrow (\exists y)(Kxy \ \& \ Ry)].$$

Since $(p \ \& \ q)$ is logically equivalent to $(q \ \& \ p)$ we could also write

$$(x)[Sx \rightarrow (\exists y)(Ry \ \& \ Kxy)].$$

Go to the next frame.

205

Some philosophers like no one who follows Hegel.

Px: x is a philosopher.
Lxy: x likes y.
Ex: x follows Hegel.

Circle the correct symbolization of the above:

1. $(\exists x)[Px \,\&\, \sim(y)(Lxy \,\&\, Ey)]$

2. $(\exists x)[Px \,\&\, (y)(Ey \rightarrow \sim Lxy)]$

3. $(\exists x)[Px \rightarrow \sim(\exists y)(Ey \,\&\, Lxy)]$

4. $(x)[Px \rightarrow (y)\sim(Ey \,\&\, Lxy)]$

---

2. $(\exists x)[Px \,\&\, (y)(Ey \rightarrow \sim Lxy)]$

If you are incorrect, go to frame 206.

If you are correct, go to frame 207.

206

'Some philosophers like no one who follows Hegel' is paraphrased as

> *There are some x* who are philosophers *and* who like no one who follows Hegel.

This in turn requires us to paraphrase

> *x* likes no one who follows Hegel.

One reading of this is

> For every follower of Hegel, *x* does not like him.

More explicitly, we write this as

> For every *y*, if *y* follows Hegel, then *x* does not like *y*.

Altogether,

> There is some *x* such that [*x* is a philosopher *and* for every *y*, (if *y* follows Hegel, then *x* does not like *y*.)]

This is symbolized by '$(\exists x)[Px \ \& \ (y)(Ey \rightarrow \sim Lxy)]$'.

Go to the next frame.

207

Use the following abbreviations:

Mx: x is male.
Ex: x is female.
Pxy: x is a parent of y.

Circle below the family relationship between k and j asserted by the following sentence:

$$(\exists x)(\exists y)[(Pxj \ \& \ Pyx \ \& \ Pyk) \ \& \ Ej \ \& \ Mk \ \& \sim(x = k)].$$

Sketching a family tree might help you.

1. j is a brother of k.

2. k is the father of j.

3. j is a granddaughter of k.

4. k is an uncle of j.

---

4. k is an uncle of j.

If you are incorrect, go to frame 208.

If you are correct, go to frame 209.

208

'$(\exists x)(\exists y)[(Pxj \ \& \ Pyx \ \& \ Pyk) \ \& \ Ej \ \& \ Mk \ \& \sim(x = k)]$, becomes interpreted as

> There is an x and a y such that x is a parent of j and y is a parent of x and y is also a parent of k . . .

Just this much tells us that k has a same parent as a parent of j. That is, k is a brother or sister of one of the parents of j. (k is not a parent of j because of the final conjunct.)

> . . . and j is female and k is male.

The information that j is female is irrelevant to this relationship. But now we know that k is a brother of one of the parents of j. Therefore k is an uncle of j.

Go to the next frame.

209

We have taken *sentential functions* and have
(1) substituted names for variables
or (2) quantified them
in order to construct *sentences*.

It is also possible to construct from *sentential functions*
(1) names of classes
and (2) names of individuals.
Instead of asserting that Herman has the property of being a dog, we might want to say

1. Herman is a member of the class of dogs.

Using a sentential function we symbolize this as

h is a member of $\hat{x}Dx$.

2. Caesar is a member of the class of noble Romans.

c is a member of $\hat{x}(Nx \mathbin{\&} Rx)$

Symbolize

3. John is a member of the class of those who, if they play, they study hard.
   j: John
   Px: x plays.
   Sx: x studies hard.

---

3. j is a member of $\hat{x}(Px \rightarrow Sx)$.

Go to the next frame.

210

We can assert that two is less than three by our usual method:

$L23$ or $2 < 3$.

Or we can use the notation of classes:

(a) The pair (2,3) is a member of the class of *ordered* pairs such that the first is less than the second.

(b) (2,3) is a member of $\hat{x}\hat{y}(x < y)$.

Notice also that we could say that

(a) 2 is a member of the class of things less than 3.

(b) 2 is a member of $\hat{x}(x < 3)$.

If we abbreviate 'is a member of' in the usual way by 'ϵ' we can symbolize

All members of the class of Greeks are members of the class of mortal things

as '$(y)(y \epsilon \hat{x}Rx \rightarrow y \epsilon \hat{x}Mx)$'.

Symbolize

The class of beautiful mermaids has no members.
Bx: x is beautiful.
Mx: x is a mermaid.

$\sim(\exists y)(y \epsilon$ ______________________

$\sim(\exists y)[y \epsilon \hat{x}(Bx \ \& \ Mx)]$ or $\sim(\exists y)(y \epsilon \hat{x}Bx \ \& \ y \epsilon \hat{x}Mx)$

Go to the next frame.

211

We write '*k* is a member of the class of those *x* such that *Bx*' as

$k \epsilon \hat{x} Bx$.

Some authors *abstract* the name of a class from a sentential function using slightly different notation. For them,

$\hat{x} Bx$ is symbolized $\{x | Bx\}$.

Thus they would write '$k \epsilon \{x | Bx\}$' for '*k* is a member of the class of *x* such that *Bx*'.

What is the meaning of '$\sim(k \epsilon \hat{x} Bx)$' (sometimes written '$k \notin \hat{x} Bx$')?

____

____

---

k is not a member of the class of those *x* such that *Bx*.

Go to the next frame.

212

Fx is a sentential function.

$(\exists x)Fx$ and $(x)Fx$ are sentences, and are either true or false.

$\hat{x}Fx$ (or: $\{x \mid Fx\}$) is a class, and is neither true nor false.

To say that something is a member of a class is to assert a sentence and thus to assert what is true or false.

$(\exists y)(y \epsilon \hat{x}Fx)$ is either true or false. It is also logically equivalent to

$(\exists y)Fy$ (and also to $(\exists x)Fx$).

$\sim(y)(y \epsilon \hat{x}Fx)$ is logically equivalent to three of the following sentences. Circle the one it is *not* equivalent to.

1. $\sim(y)Fy$

2. $(y)(y \notin \hat{x}Fx)$

3. $(\exists y)(y \notin \hat{x}Fx)$

4. $\sim(x)(x \epsilon \hat{x}Fx)$

---

2. $(y)(y \notin \hat{x}Fx)$ This asserts that nothing is a member of the class denoted by '$\hat{x}Fx$'.

Go to the next frame.

213

There are various assertions involving classes that one can make. A certain individual—say, k—is a member of the class: $k \epsilon \{x \mid Fx\}$. Something is a member of the class: $(\exists y)(y \epsilon \hat{x}Fx)$.

One can also say that one class is a member of another class:

$$\hat{x}Fx \epsilon \hat{y}Gy.$$

Or one can say a class is a *subclass* of another class—$\hat{y}Gy \subset \hat{z}Hz$—or even that one class is *identical* to another: $\hat{x}Rx \quad \hat{y}Sy$.

Put the correct sign between the names of the above two classes.

The identity sign, '='—therefore, $\hat{x}Rx = \hat{y}Sy$.

Go to the next frame.

214

A class is characterized by its members. Two classes are identical if and only if they have exactly the same members.

$(\hat{x}Rx = \hat{y}Sy) \leftrightarrow \hat{x}Rx$ *has exactly the same members as* $\hat{y}Sy$.

Now the problem is to symbolize the right side of the above biconditional sentence. Assuming the classes are identical, suppose $z\epsilon\hat{x}Rx$; then z (is/is not) a member of $\hat{y}Sy$. (Choose one.)

---

z is a member of $\hat{y}Sy$.

Suppose further that $z\notin\hat{x}Rx$; then z (is/is not) a member of $\hat{y}Sy$.

$z\notin\hat{y}Sy$

So the right side of the biconditional is

For any z, if $z\epsilon\hat{x}Rx$ then $z\epsilon\hat{y}Sy$, and if $z\notin\hat{x}Rx$ then $z\notin\hat{y}Sy$.

Symbolically,

$(z)[(z\epsilon\hat{x}Rx \rightarrow z\epsilon\hat{y}Sy)$ & $(z\notin\hat{x}Rx \rightarrow z\notin\hat{y}Sy)]$.

Looking carefully at this formula, or thinking about the conditions for identity, might lead one to reformulate the right side of the biconditional as

For any z, $z\epsilon\hat{x}Rx$ if and only if $z\epsilon\hat{y}Sy$.

$(z)(z\epsilon\hat{x}Rx \leftrightarrow z\epsilon\hat{y}Sy)$

Go to the next frame.

215

You have seen some of the versatility of the notation of symbolic logic. Beginning with a

sentential function: Fx

we can symbolize two kinds of sentences:

singular sentence: Fa.

quantified sentence: $(x)Fx$ and $(\exists x)Fx$.

We can also symbolize names, or expressions that denote (which are neither true nor false):

class name: $\hat{x}Fx$ or $\{x \mid Fx\}$

Recalling that two classes are identical if they have the very same members, consider the following problem:

If $b \epsilon \{x \mid Ax\}$ and $j \epsilon \{y \mid Cy\}$ and $\{x \mid Ax\} = \{y \mid Cy\}$, is $b = j$?

Not necessarily. There is a member of $\{y \mid Cy\}$ that is identical to b, but j need not be that particular individual.

Go to the next page.

Here is a summary of the types of symbolizations you have learned.

1. Dante is a cat who smiles. $Cd \,\&\, Sd$

2. There is a cat who smiles. $(\exists x)(Cx \,\&\, Sx)$

3. All cats smile. $(x)(Cx \rightarrow Sx)$

4. No cats smile. $(x)(Cx \rightarrow \sim Sx)$ or $\sim(\exists x)(Cx \,\&\, Sx)$

5. Some cats do not smile. $(\exists x)(Cx \,\&\sim Sx)$

6. The class of cats who smile $\hat{x}(Cx \,\&\, Sx)$ or $\{x \mid (Cx \,\&\, Sx)\}$

7. There are exactly two cats who smile.
$$(\exists x)(\exists y)\big((Cx \,\&\, Sx) \,\&\, (Cy \,\&\, Sy) \,\&\sim(x = y) \,\&\, (z)\{(Cz \,\&\, Sz) \rightarrow [(x = z) \vee (y = z)]\}\big)$$

Go to the next page.

Part II Review with Comments

(1) You should be able to recognize when a formula correctly symbolizes an English sentence.

Comment: Naturally you will have a hard time puzzling out such brain teasers as

Nobody loves anybody

but you should not have too much trouble recognizing the correct symbolization of

Some animals have no toes.

(2) You should also be able to recognize when an English sentence correctly interprets a formula.

Comment: Always seek out truth conditions. The analysis of the truth conditions of a formula is unambiguous and precise. Unfortunately, analyses of English sentences are not so straightforward, since they often suffer from vagueness and ambiguity.

(3) You should be able to figure out a few simple logical equivalences or contradictions between formulas.

Comment: You have only two tools for this. One, you know how to move a negation sign back and forth through quantifiers, and two, you can replace logically equivalent sentential functions within a formula.

(4) You should be able to recognize the *classes* denoted by the relevant symbolic expressions.

Comment: When the sentential function involved is not too complex, the task is easy:

$\hat{x}(Ax \vee Bx)$

The set of those things that are A or are B.

But when the sentential function following the *abstraction operator* is complicated, things are not so easy.

$$\hat{x}(y)(z)\{[(y > 1 \ \& \ z > 1 \ \& \ y < x \ \& \ z < x) \rightarrow y \cdot z \neq x] \ \& \ 20 < x \ \& \ x < 30\}$$

is the class of prime numbers (numbers evenly divisible only by one and themselves) between 20 and 30. This class is identical with the class consisting of 23 and 29—that is, $\hat{x}(x = 23 \vee x = 29)$.

Go to the next page.

Self-Examination

1. Select the formula that correctly symbolizes the English sentence

   Some Republicans vote against any bill introduced by a Democrat.

   Rx: x is a Republican.
   Bx: x is a bill.
   Dx: x is a Democrat.
   Vxy: x votes against y.
   Ixy: x introduced y.

   (a) $(x)\{(Rx \ \& \ Vxy) \rightarrow (y)[By \ \& \ (\exists z)(Dz \ \& \ Izy)]\}$

   (b) $(\exists x)\{Rx \ \& \ (y)[(Vxy \ \& \ By) \rightarrow (\exists z)(Dz \ \& \ Iyz)]\}$

   (c) $(\exists x)(Rx \ \& \ (y)\{[By \ \& \ (\exists z)(Dz \ \& \ Izy)] \rightarrow Vxy\})$

   (d) $(x)\{Rx \rightarrow (y)[(Vxy \rightarrow By) \ \& \ (z)(Dz \rightarrow Izy)]\}$

2. Select the English sentence that correctly interprets the formula

   $(x)(y)[Px \rightarrow ({\sim}Kxy \rightarrow {\sim}Uyx)]$

   Px: x is a person.
   Kxy: x knows y.
   Uxy: x hurts y.

   (a) A person doesn't always know whom he hurts.

   (b) One always hurts someone he knows.

   (c) Some people are hurt if they don't know something.

   (d) What a person doesn't know won't hurt him.

3. Which formula symbolizes a contradiction to '$(x)(Px \rightarrow Sx)$'?

   (a) $(x) \sim (Px \rightarrow Sx)$ (c) $(\exists x) \sim (Px \ \& \ Sx)$

   (b) $(x)(Px \rightarrow {\sim}Sx)$ (d) $(\exists x)(Px \ \& {\sim}Sx)$

4. '${\sim}(\exists x)(Tx \vee Mx)$' and '$(x)({\sim}Tx \ \& {\sim}Mx)$' are

   (a) logically equivalent

   (b) contradictory

   (c) neither equivalent nor contradictory

   (d) indeterminate until 'T' and 'M' are interpreted

5. If 'John is a bad boy' is symbolized 'Aj & Bj', then what does '$\hat{x}(Ax \& Bx)$' symbolize?

   (a) All boys are bad.

   (b) Some boys are bad.

   (c) The class of bad boys

   (d) x is bad and x is a boy.

6. Interpret '$(x)[Fxs \rightarrow (x = t \vee x = d \vee x = h)]$'.

   Fxy: x can fix y.
   s: the sink
   t: Tom
   d: Dick
   h: Harry

   (a) Only Tom, Dick, or Harry can fix the sink.

   (b) Any Tom, Dick, or Harry can fix the sink.

   (c) Tom can fix the sink, or Dick can fix the sink, or Harry can fix the sink.

   (d) The sink can be fixed by Tom, Dick, or Harry.

7. In the following, what relationship does j have to k?

   $$(\exists x)(\exists y)(\exists z)[Pxk \& Pyx \& Pzx \& Pyj \& Pzj \& Mx \& Ej \& \sim(y = z) \& \sim(x = j)]$$

   Pxy: x is a parent of y.
   Mx: x is male.
   Ex: x is female.

   (a) j is a paternal aunt of k.

   (b) j is a niece of k.

   (c) j is a paternal grandmother of k.

   (d) j is a first cousin of k.

ANSWERS

1. c
2. d
3. d
4. a
5. c
6. a
7. a

End of Part II

Part 3

Ordering Relations and Their Properties

Introduction

The validity of some arguments involving relations depends on certain structural features of these relations. From the fact, for instance, that some event was earlier than another, we might infer some conclusion based on some features of the relation of being earlier than. We should be able to recognize and to state these features explicitly. This is important because arguments in a natural language often conceal these properties, and we need to state them when we formally reconstruct the argument. In addition, the special properties of some relations enable them to be used to order the elements of a set. Ordering relations are basic to any theory of measurement, and measurement is a concept closely allied with the notion of objectivity in science. For these reasons, Part III extends what we have learned to ordering relations and properties of relations in general.

Upon completion of this section, you should be able to identify such important kinds of orderings as strict simple, strict partial, simple, partial, and weak. You should be able to inspect a relation and determine its transitivity, symmetry, reflexivity, and connectivity properties. You will learn what an equivalence relation is, and you should be able to recognize one. Finally, you should be able to recognize the complement and the converse of a relation.

TEAR THIS SHEET OUT

-------------------------- fold here --------------------------

216

One important part of reasoning, in ordinary affairs as well as in scientific investigations, is making comparisons. We might, for instance, judge that John is taller than Mary, and that Mary is taller than Bill. These comparisons are based on a relationship of height, and the phrase '. . . is taller than . . .' expresses that *relation*.

We might also claim that

Bill is taller than no one.

This is symbolized by which of the following?

1. $\sim(x)Tbx$

2. $(x)\sim Tbx$

3. $(\exists x)\sim Tbx$

4. $\sim(\exists x)Tbx$

If you picked choice 1, go to frame 217.

2, go to frame 218.

3, go to frame 219.

4, go to frame 220.

217

'$(x)Tbx$' symbolizes 'Bill is taller than everyone'. The negation merely denies this: 'It is not the case that Bill is taller than everyone'.

You may wish to review the material beginning on frame 133 before you return to frame 216 to try again.

218

'$(x) \sim Tbx$' is correct. You might look back at frame 216 to notice that '$\sim(\exists x)Tbx$' is also correct.

Turn to frame 221.

219

Since Bill is not taller than himself, there is someone whom Bill is not taller than. So '$(\exists x){\sim}Tbx$' is true, but it does not symbolize the sentence in question.

Return to frame 216 and try again.

220

'$\sim(\exists x)Tbx$' is correct. You might look back at frame 216 to notice that '$(x) \sim Tbx$' is also correct.

Go to the next frame.

221

We have until now silently agreed to talk only about persons, but strictly and explicitly,

$$(x)\sim Tbx \qquad \text{or} \qquad \sim(\exists x)Tbx$$

translates as 'Bill is taller than *nothing*'—not taller than my desk (unlikely); not taller than my typewriter (false); not taller than the square root of five (not clearly sensible).

In such cases, the collection of items among which comparisons are being made must be specified or made clear by the context. Determining the truth of a sentence that asserts a relation between two *named* individuals is not a problem, but when the sentences are quantified, problems exist. After all, Bill may just be taller than no *one in this room*, although not taller than some persons outside the room. Or he may just be taller than no *student in the school.* Or he may be taller than no *living person.* And so on.

If room 202 is completely empty, the sentence

Bill is taller than no one in room 202

is (true/false/unknown).

If you say it is true, go to frame 222.

false, go to frame 223.

unknown, go to frame 224.

222

Correct. Restricting the variable 'x' to persons in room 202, '$(\exists x)Tbx$' is false. Hence, '$\sim(\exists x)Tbx$' is true.

Go to frame 225.

223

If 'Bill is taller than no one in room 202' is false, then 'Bill is taller than someone in room 202' is true. How can this be, if room 202 is empty?

Return to frame 221.

224

If we do not restrict the variable 'x' to persons in room 202, and explicitly symbolize the sentence with these abbreviations

Txy: x is taller than y
Rx: x is a person in room 202

we obtain the formula

$\sim(\exists x)(Rx \;\&\; Tbx)$.

This is the denial of the sentence

Something is a person in room 202 *and* Bill is taller than that person.

The truth-value of this sentence is easily established, given that room 202 is empty. With this hint, return to frame 221.

225

Confining our attention to students currently registered at a school, we could compare and rank them according to height, weight, or age. We could even *order* students alphabetically, or according to their student identification numbers.

We say that we *order* a collection of items alphabetically when we use the *relation* of *being alphabetically earlier than*, so that for any two adjacent items in the ordering, the first is alphabetically earlier than the second.

Is there a relation that orders all the individuals in this collection (the students currently registered at some school) by their IQ test scores?

Yes / No

If you say yes, go to frame 226.

no, go to frame 227.

226

Only if all students have taken this test and received a score on it, or if we cleverly devise a complicated relation to take care of those students with no score (which is not the same as having a score of zero).

Go to the next frame.

227

If some students have not taken an IQ test, then there is no score for them. Suppose David has no IQ test score and Albert does. It is difficult to say how to compare David and Albert with respect to IQ test scores, although there are many artificial ways to solve this problem—for instance, assign individuals with no score a score of zero.

> When every pair of *distinct* individuals in a collection is comparable, one way or the other, by a specific relation, we say that relation is *connected* in that collection (or class, or set).

In our symbolic language, a relation R is connected in a set S if for every x and y in the set S,

$$(x \neq y) \rightarrow (Rxy \vee Ryx).$$

One must be careful to investigate the individuals in the set S as well as the relation R. For instance, if there are two different individuals in the set of all registered students at a school who are exactly the same height, then the relation of being taller than is *not* connected.

For the set of all students at a school, is the following relation connected?

Rxy: x is at least as heavy as y.

Yes / No

If you say yes, go to frame 228.

no, go to frame 229.

228

Correct. For any two different individuals, one is as heavy or heavier than the other, or vice versa. Notice that they don't even have to be different individuals, since everyone is as heavy as himself or herself. This point will be discussed further later on in this section.

Go to frame 230.

229

Sorry, but the relation *is* connected in that set.

Reread frame 227 and choose the correct answer.

230

It is often important to know whether any two arbitrary and distinct individuals in the set you are interested in can be compared one way or the other. Thus, connectedness of a relation in a set is often an important property. Some of the relations we shall examine will be connected (in a set), and others will not. But *all* the relations that give some kind of order to a set have the property of being *transitive*.

What does it mean for a relation to be transitive in a set? Recall the earlier example in which we judged that John was taller than Mary, and that Mary was taller than Bill. Without looking at any of the people involved, we know that John is taller than Bill.

'Being taller than' expresses a transitive relation (among physical objects). This is symbolized

$$(x)(y)(z)[(Txy \ \& \ Tyz) \rightarrow Txz].$$

Does 'being a next-door neighbor of' express a transitive relation?

Yes / No

If you said yes, go to frame 231.

no, go to frame 232.

231

Only in very unusual neighborhoods! In typical neighborhoods, the relation is not transitive. The question cannot be answered until the set is specified.

Go to frame 233.

232

The relation is not transitive in a typical neighborhood. If one had no neighbors, on the other hand, the antecedent condition of the description of transitivity would not be satisfied in this set, and so the relation would be (vacuously) transitive. The question cannot be answered until the set is specified.

Go to the next frame.

233

That was a little unfair, but people frequently presuppose or quietly assume a typical or well-behaved set when they attribute a property to a relation.

You may relax your guard now and consider the set of human beings. Which of the following relations is transitive in that set?

1. x is the father of y.

2. x is a full-blooded brother of y.

3. x is a first cousin to y.

4. x is married to y.

If you chose answer 1, go to frame 234.

2, go to frame 235.

3, go to frame 236.

4, go to frame 237.

234

Answer 1 is not correct. When the antecedent condition is satisfied, the first individual will be the grandfather of the third.

Reread frame 230 and then try frame 233 again.

235

Answer 2, 'x is a full-blooded brother of y', is correct. A brother of my brother is also a brother of mine. (We have excluded step-brothers.)

Go to frame 238.

236

Answer 3 is incorrect. You can have first cousins on your mother's side as well as on your father's side. We can imagine a special family situation where this relation would be transitive, but not in the set of human beings.

Reread frame 230 before you try frame 233 again.

237

Answer 4 is incorrect. In a monogamous marriage, where two and only two distinct persons are ever related by marriage at one time, the relation expressed by '. . . is married to . . .' is not transitive. In non-nuclear extended families, the relation of being married to may possibly be transitive, but it is not in the set of all human beings.

Review frame 230 before trying frame 233 again.

238

If you are going to order the individuals in a collection, you must be sure that you use a transitive relation, because *all* orderings in sets are transitive. If you wish to order *all* the individuals, the ordering relation must be connected.

For some relations it may happen that the relation holds both ways for some pairs of individuals. That is, for a relation, R, and individuals a and b,

Rab & Rba

is true. When this occurs, we shall say that R is *reciprocal* for a and b. Thus the relation

. . . is as old as . . .

is reciprocal for some pairs of human beings, because some persons are the same age. And also because, significantly, everyone is as old as himself/herself. (Notice that we did not require a and b to be different individuals.)

Is the relation

. . . is older than . . .

reciprocal for any pairs of humans?

Yes / No

If you answer yes, go to frame 239.

no, go to frame 240.

239

Incorrect. If a is older than b, then b is not older than a.

Go to the next frame.

240

If a is older than b, then b is not older than a.

When a relation is not reciprocal for any pairs in a set, we say it is *asymmetric* in the set.

ASYMMETRY: $(x)(y)(Rxy \rightarrow {\sim}Ryx)$

In a set of automobiles registered in a state, which one of the following relations is asymmetric?

(a) . . . costs as much or more than . . .
(b) . . . has at least as high fuel consumption as . . .
(c) . . . has a lower license number than . . .
(d) . . . is no heavier than . . .

(c) . . . has a lower license number than . . .

Go to frame 241.

241

If a relation is connected, transitive, and asymmetric in a set, we have a *strict simple* ordering.

One example of a strict simple ordering you are probably familiar with is the relation of being less than (<) in the whole numbers. Convince yourself that the following three sentences are true in the collection of whole numbers (. . . , −2, −1, 0, 1, 2, 3, . . .):

1. $(x)(y)[x \neq y \rightarrow (x < y \vee y < x)]$ connected

2. $(x)(y)(z)[(x < y \;\&\; y < z) \rightarrow x < z]$ transitive

3. $(x)(y)(x < y \rightarrow {\sim} y < x)$ asymmetric

A strict simple ordering is the first ordering we shall consider. Notice that '$a < b$' and '$b < a$' cannot both be true—that is to say, the relation is not reciprocal for any pairs of individuals. And notice that all the individual numbers are ranked in comparison with all of the others.

Which of the following denotes a strict simple ordering?

(a) '. . . is an ancestor of . . .' relative to all humans
(b) '. . . costs as much or more than . . .' relative to items for sale in a given store
(c) '. . . has a higher telephone number than . . .' relative to subscribers within the same area-code district

If you picked choice a, go to frame 242.

b, go to frame 243.

c, go to frame 244.

242

. . . is an ancestor of . . . is not connected. For any two different human beings, it is not always the case that one is an ancestor of the other.

Return to frame 241 and try again.

243

. . . costs as much or more than . . . may be reciprocal. The relation is not asymmetric, because two items *a* and *b* could have the same price; hence 'Cab & Cba' would be true, and '$Cab \rightarrow \sim Cba$' would be false.

Return to frame 241 and try again.

244

Answer c is correct. Each subscriber has a number, and no two have the same number.

Lining up items in a straight line leads to a simple way of ordering them. The different points on a straight line are *simply* ordered by the relations of being to the right of, or being higher than. And because these relations are *asymmetrical,* they are *strict* simple orderings.

Think about the ancestor relation among human beings. If we were to trace through a family tree choosing only one person from each generation (say, the firstborn), then for the set of these individuals the relation denoted by 'is an ancestor of' would be a strict simple ordering.

Suppose we do this for two families—say, the Jones family and the Smith family. Now take the ancestor relation with respect to the combined collection of individuals so chosen from both families. Which condition for a strict simple ordering is no longer satisfied?

(a) connected
(b) transitivity
(c) asymmetry

(a) connected. It no longer is the case that for *any* two distinct individuals, one is the ancestor of the other. One could be a Smith and the other a Jones.

Go to the next frame.

245

What we now have is an incomplete ordering, or a *strict partial* ordering. There still are no cases where '*Rab*' and '*Rba*' both hold, but all the individuals are not on the *same* line, one after the other.

In addition, the relation denoted by 'is heavier than' gives a *strict partial* ordering to the set of students in this school, because some students weigh the same. Thus there are pairs of distinct students such that neither '*Hab*' nor '*Hba*' are true.

A relation is a strict partial ordering of a set when it has the two

properties of ____________________ and ____________________.
(Fill in the words.)

---

transitivity and asymmetry (or vice versa)

Go to the next frame.

246

ORDERINGS

Strict Simple	*Strict Partial*
transitive	transitive
asymmetric	asymmetric
connected	

These are strict orderings because the relations are never reciprocal. And this prohibition is achieved by the asymmetry condition. We are now going to allow relations to be reciprocal, and so we must look more closely at various symmetry properties.

A relation R is called asymmetric in a set just when, for every x and y in the set, $(Rxy \rightarrow \sim Ryx)$.

We shall call a relation *symmetric* in a set whenever, *if* the relation holds between two individuals one way, it also holds the other way. That is, whenever the relation holds, it is reciprocal.

$(x)(y)(Rxy \rightarrow Ryx)$ where the quantifiers range over the set in question.

Identity is an obvious symmetric relation. Which of the following are *not* symmetric in the indicated sets of individuals?

(a) 'is a brother of' {human beings}
(b) 'is as tall as' {buildings in New York}
(c) 'is a neighbor of' {residents of Boston}
(d) 'is the square root of' {Natural numbers: 0, 1, 2, 3, . . .}

Answer (a) because of sisters, and answer (d) because 4 is not the square root of 2. Warning: What if the set in answer (d) were {0, 1}?

Go to the next frame.

247

An ordering by an asymmetric relation is a strict ordering. An asymmetric relation is *never* reciprocal for individuals.

On the other hand, a symmetric relation is *always* reciprocal for those individuals satisfying the relation. We want now to look at relations that are reciprocal some of the time.

Specifically, we shall look at relations that are reciprocal *only when* the individuals are identical. In symbols, for any x and y in the set,

$$(Rxy \ \& \ Ryx) \rightarrow x = y.$$

Such a relation is called *anti-symmetric*.

A nice example of an anti-symmetric relation is that of being less than or equal to in the set of whole numbers. This relation is not asymmetric, and it is not symmetric. It is anti-symmetric because

$$(a \leq b \ \& \ b \leq a) \rightarrow a = b.$$

Which *one* of the following ordering relations is anti-symmetric?

(a) 'is no heavier than'	{human beings}
(b) 'is the same or an earlier date than'	{dates on a calendar}
(c) 'is at least as old as'	{human beings}
(d) 'took place at the same time or earlier than'	{historical events}

(b) 'is the same or an earlier date than' 'a is the same or an earlier date than b and b is the same or an earlier date than a' could only be true if a and b are the same date.

Go to the next frame.

248

If a relation is symmetric, it does not help us to order things in a line because it is reciprocal for all pairs for which the relation holds. But an anti-symmetric relation can help in ordering individuals, at least a bit, because it permits the relation to hold only one way for at least some individuals in the set.

Take the relation expressed by 'is the square root of'. In the set of natural numbers greater than one, this relation is asymmetric. (2 is the square root of 4, 3 is the square root of 9, etc.)

But we have noted that 0 is the square root of 0, and 1 is the square root of 1. We can see that, in the set of all natural numbers, if Sxy & Syx, then $x = y$. In this set the relation is anti-symmetric and not asymmetric. (This illustrates, incidentally, the importance of specifying the set of individuals being related.)

In the set of natural numbers, there are *some* individuals (namely 0 and 1) for which

x is the square root of x

is true, although for most natural numbers the above expression is not true. A relation that *every* individual in the set has to itself is called *reflexive*.

A relation R is reflexive in a set if

1. $(x)(y)(Rxy \vee Ryx)$

2. $(x)(y)(Rxy \rightarrow Ryx)$

3. $(x)(\exists y)Rxy$

4. $(x)Rxx$

If you chose 1, go to frame 249.

2, go to frame 250.

3, go to frame 251.

4, go to frame 252.

249

Choice 1 is incorrect. If the set contains only one member, then '$(x)(y)(Rxy \vee Ryx)$' is equivalent to the correct choice.

Review frame 248 and try again.

250

Choice 2 is incorrect. This formula expresses the symmetry property.

Review frame 248 and try again.

251

Choice 3 is wrong. '$(x)(\exists y)Rxy$' asserts merely that every x bears the relation R to *something or other*. We want the condition that every x bears R to *itself*.

Go to the next frame.

252

Choice 4, '$(x)Rxx$', asserts that every x bears R to itself, the defining property of a reflexive relation.

We shall have to be alert to such facts as that no one *is his own brother*, that some people *love* themselves, and that everyone *is as old as himself*. In symbols,

$(x) \sim Bxx$
$(\exists x)Lxx$
$(x)Oxx$

Not every relation is reflexive.

Go to the next frame.

253

We are now going to take a big step, so let us proceed very carefully and explicitly. We are going to consider ordering relations that are *not* asymmetric, but that are anti-symmetric and reflexive instead. So we drop the adjective 'strict' from their titles. Schematically this can be represented as follows:

Strict Partial		*Partial*
transitive	→	transitive
asymmetric	→	anti-symmetric
		reflexive

Complete the pattern below:

Strict Simple		*Simple*
transitive	→	________________
connected		________________
asymmetric	→	________________
		________________

Simple
transitive
connected
anti-symmetric
reflexive

Go to the next frame.

254

Partial and simple orderings, either strict or not, are the most common orderings. Actually, there is an ordering that is weaker than any of these four. If a relation is (1) transitive in a set, and (2) holds one way or the other between *any* two individuals in the set, it is a *weak* ordering.

Weak orderings are more liberal than either partial or simple orderings. They permit "ties." Two or more distinct individuals are tied in an ordering if they are ordered with respect to *other* individuals in the set, but not between, or among, themselves. That is, the ordering relation does not discriminate among tied individuals.

Suppose David and Mark weigh the same. Then the relation expressed by 'is at least as heavy as' is an ordering (of the students in a class) in which David and Mark are tied. Notice that being at least as heavy is not anti-symmetric, for although 'Hdm & Hmd', is true, David is not identical to Mark. This is a weak ordering.

Let us symbolize the properties of a weak ordering:

1. $(x)(y)(z)[(Rxy \,\&\, Ryz) \rightarrow Rxz]$
2. $(x)(y)(Rxy \vee Ryx)$

Formula 1 is the condition of transitivity. Is formula 2 the condition of being connected?

Yes / No

If your answer is yes, go to frame 255.

no, go to frame 256.

255

A relation R is connected in a set A if for any two *distinct* individuals x and y in A, $(Rxy \vee Ryx)$. We have omitted the condition that the individuals be distinct. If this distinction slipped your mind, you might review frame 227.

Go to the next frame.

256

A relation R is connected if $(x)(y)[x \neq y \rightarrow (Rxy \vee Ryx)]$. The condition $(x)(y)(Rxy \vee Ryx)$ is *not* the condition of being connected; it is a stronger condition—namely, that of being *strongly connected*.

A relation that is transitive and *strongly connected* in a set is a weak ordering of the set.

If an ordering relation R is reciprocal for a and b—that is, if $(Rab \ \& \ Rba)$—then

(i) if R is a *partial* or *simple* ordering, a and b are ________

____

(ii) if R is a *weak* ordering, a and b may be ________________

____

(Fill in the above blanks.)

(i) if R is a partial or simple ordering, a and b are *identical OR the same individual*.

(ii) if R is a weak ordering, a and b may be *nonidentical OR different individuals*.

Go to the next frame.

257

The properties of relations we have used so far are

1. transitive
2. asymmetric
3. anti-symmetric
4. connected
5. strongly connected
6. reflexive

Place the appropriate numbers of these properties after the orderings they belong to.

(a) strict partial 1, ________________

(b) strict simple 1, ________________

(c) partial 1, ________________

(d) simple 1, ________________

(e) weak 1, ________________

(I have started you out with transitivity.)

(a) 1, 2 (Review frames 244 and 245.)
(b) 1, 2, 4 (Review frame 241.)
(c) 1, 3, 6 (Review frame 253.)
(d) 1, 3, 4, 6 (Review frame 253.)
(e) 1, 5 (Review frame 254.)

Go to the next frame.

258

We have been concentrating on relations that are transitive in a set. What about relations that are not transitive?

$$(x)(y)(z)[(Rxy \ \& \ Ryz) \rightarrow Rxz]$$

being false means that some instance

$$(Rab \ \& \ Rbc) \rightarrow Rac$$

is false, and so the sentence

$$(Rab \ \& \ Rbc) \ \& \sim Rac$$

is true.

If *all* such instances are true, then it is true that

$$(x)(y)(z)[(Rxy \ \& \ Ryz) \rightarrow \sim Rxz].$$

This is the property of being *intransitive*. In words, whenever three (not necessarily different) individuals satisfy '$(Rxy \ \& \ Ryz)$', then 'Rxz' is false. Which of the following are intransitive?

(a)	brother of	{human beings}
(b)	costs as much as	{items in Bloomingdale's}
(c)	father of	{human beings}
(d)	square root of	{natural numbers: 0, 1, 2, 3, 4, . . . }

Answer (c), father of, is intransitive. Answer (b) is transitive, and answers (a) and (d) are neither transitive nor intransitive. [For $Sxy : x$ is the square root of y, '$(S11 \ \& \ S11) \rightarrow \sim S11$' is fale, so S is not intransitive.]

Go to the next frame.

259

So with respect to transitivity, there are three possibilities:

1. transitive

2. intransitive

3. neither transitive nor intransitive

We have used the properties of asymmetry and anti-symmetry and have briefly mentioned the property of being symmetric. Please match the name of the property with its symbolic formulation:

(a) asymmetric	1. $(x)(y)[(Rxy \,\&\, Ryx) \rightarrow x = y]$
(b) anti-symmetric	2. $(x)(y)(Rxy \rightarrow Ryx)$
(c) symmetric	3. $(x)(y)(Rxy \rightarrow \sim Ryx)$

(a) asymmetric	3. $(x)(y)(Rxy \rightarrow \sim Ryx)$
(b) anti-symmetric	1. $(x)(y)[(Rxy \,\&\, Ryx) \rightarrow x = y]$
(c) symmetric	2. $(x)(y)(Rxy \rightarrow Ryx)$

Go to the next frame.

260

With regard to symmetry properties, it is possible for a relation to be neither symmetric, nor asymmetric, nor anti-symmetric.

Using the proper letters, identify the symmetry properties of the listed relations:

(a) symmetric
(b) asymmetric
(c) anti-symmetric
(d) neither (a) nor (b) nor (c)

1. father of	{humans}	________
2. lover of	{humans}	________
3. less than or equal	{whole numbers}	________
4. spouse of (married to)	{humans}	________
5. more nutritious than	{breakfast cereals}	________

---

1. b
2. d
3. c
4. a
5. b

If you need to review, see frames 246 and 248.

Go to the next frame.

261

Finally, we want to see what possibilities there are when a relation is *not* reflexive in a set. If a relation R is reflexive, this means

$$(x)Rxx.$$

For example, in the set of human beings, everyone weighs as much as himself or herself. But no one weighs more than himself or herself. Thus

$$(x)\sim Rxx.$$

A relation with this property is called *irreflexive* in the set.

There are then three possible reflexivity properties of a relation in a set:

(a) reflexive
(b) irreflexive
(c) neither reflexive nor irreflexive

Use the letters a, b, and c to indicate the appropriate reflexive properties of the following relations:

1. identical {everything} __________
2. in love with {humans} __________
3. cube root of {natural numbers} __________
4. pays as many taxes as {U.S. citizens} __________
5. mother of {humans} __________

---

1. a
2. c (Some people do, you know.)
3. c (Don't forget 0 and 1.)
4. a
5. b

Go to the next frame.

262

We can now take a relation in the set of whole numbers, such as being less than or equal to, and determine its properties: it is transitive, anti-symmetric, reflexive, and connected (even strongly connected). Hence this relation is a simple ordering of the whole numbers.

What about the relation expressed by 'ancestor of' in the collection of all (living and dead) human beings? This relation is

1. transitive/intransitive/neither

2. symmetric/anti-symmetric/asymmetric/none

3. reflexive/irreflexive/neither

4. strongly connected/connected/neither
   (Circle the correct words.)

---

1. transitive
2. asymmetric
3. irreflexive
4. neither

This relation is a ____________________ ordering of the set of human beings.

---

strict partial (If necessary, review frames 244 and 245.)

Go to the next frame.

263

If John loves Mary, then Mary is loved by John. If Brutus stabbed Caesar, then Caesar was stabbed by Brutus. In general, if a relation holds one way between two individuals, then there is a relation that holds between the individuals taken in the *reverse order*. This second relation is called the *converse* of the first relation.

When a relation is reciprocal for individuals a and b, the *same* relation holds both ways—that is, Rab and Rba. The converse of a relation is usually (but not always) a *different* relation.

'Is loved by' expresses the converse of the relation of loving. 'Being stabbed by' expresses the converse of the relation of stabbing.

The converse of R is often written as '$\breve{R}$'. So

$$(x)(y)(\breve{R}xy \leftrightarrow Ryx).$$

'Descendant of' expresses the converse of ____________________

---

ancestor of

Go to the next frame.

264

Some other relations are

Relation	*Converse Relation*
parent of	________________
classmate of	________________
________________	identical to

(Fill in the blanks.)

---

parent of	child of
classmate of	classmate of
identical to	identical to

Some relations, we see, are their own converses.

The converse of the relation expressed by 'less than' is the relation called 'greater than'. But the converse of the relation written 'less than or equal to' ($\leq$) is ____________________________________
(Fill in.)

---

greater than or equal to ($\geq$).

Go to the next frame.

265

The converse of a relation is not the same as the *complement* of a relation. The complement of a relation R, often written '$\bar{R}$', holds for all those pairs of individuals in the set for which R does *not* hold.

Thus, the complement of '=' is '≠'. The complement of '<' is '≥'. The complement of 'being stabbed by' is 'not being stabbed by'.

Fill in the blank spaces below. For human beings

Relation, R	Converse, $\breve{R}$	Complement, $\bar{R}$
older than	younger than	younger than or as old
as heavy or heavier than	________	lighter than
born in the same country as	________	________
________	________	not a parent of

---

as heavy or heavier than	lighter than or as light as	lighter than
born in the same country as	born in the same country as	not born in the same country as
a parent of	a child of	not a parent of

Go to the next frame.

266

If a relation is an ordering of a set, so is its converse. In fact, the converse gives exactly the same kind of ordering; it merely reverses the order, preserving any ties there are.

If a relation is an ordering of a set, is its complement, in general, also an ordering of the set? You may wish to examine the examples from the previous frame.

Yes / No

If you answered yes, go to frame 267.

no, go to frame 268.

267

You said the complement of an ordering relation is an ordering. This is not correct. Frequently the complement of a transitive relation is not itself transitive and hence could not be any kind of ordering relation. For example, the complement of 'is a classmate of' is not transitive.

Review frames 263 and following.

268

Correct. Often the complement of a transitive relation is not even transitive, and so could not be an ordering relation.

Go to the next frame.

269

Any relation that is transitive, symmetric, and reflexive in a set is called an *equivalence* relation. Equivalence relations do not order the set. Instead, they categorize or classify the individuals into separate compartments where all the individuals in any one compartment, or partition, are treated alike.

Thus, for human beings, 'born in the same year as' does not order individuals in some ranking, but it does classify individuals into distinct classes. Each individual in any class was born in the same year as each individual in that class. 'Born in the same year as' is reflexive, symmetric, and transitive in the set of human beings; it is an *equivalence relation*.

Which of the following are *not* equivalence relations?

1. 'is identical to' {everything}
2. 'has the same parents as' {human beings}
3. 'is less than or equal to' {whole numbers}
4. 'is married to' {human beings}

---

3. 'is less than or equal to' is an ordering relation.
4. 'is married to' is symmetric but not reflexive or transitive.

Self-Examination

A consumer's organization assigned overall numerical ratings to fifteen models of 35 mm cameras as follows:

Model	*Rating*	*Model*	*Rating*
One	72	Nine	66
Two	71	Ten	65
Three	71	Eleven	64
Four	70	Twelve	62
Five	69	Thirteen	60
Six	69	Fourteen	59
Seven	69	Fifteen	56
Eight	67		

Use these labels for the first four questions:

(a) strict simple ordering
(b) strict partial ordering
(c) simple ordering
(d) partial ordering
(e) weak ordering
(f) equivalence relation

1. In the set of *all* models of 35 mm cameras, what is the relation 'is rated numerically higher than'? ___________

2. In the set of the fifteen models tested and rated, what is the relation 'is not rated numerically higher than'? ___________

3. In the set of the fifteen models tested and rated, what is the relation 'has the same numerical rating as'? ___________

4. If models Three, Five, and Six are eliminated from the above set, what is the relation 'is rated numerically higher than' for the remaining twelve models? ___________

5. Which of the following formulas expresses the fact that R is *transitive* in a set A, where for any x and y in A

   (a) $(Rxy \mathbin{\&} Ryx) \rightarrow Rxx$
   (b) $(Rxy \mathbin{\&} Rxz) \rightarrow Ryz$
   (c) $(Rxy \mathbin{\&} Ryz) \rightarrow Rxz$
   (d) $(Rxy \mathbin{\&} Ryz) \rightarrow \sim Rxz$

6. If for any x and y in a set A, $(Rxy \rightarrow \sim Ryx)$, the relation R is
   (a) symmetric
   (b) asymmetric
   (c) anti-symmetric
   (d) none of these

7. The converse, $\breve{R}$, of a relation among kinds of minerals is expressed by 'as hard or harder than'. The original relation, R, is

____

   (a) as hard or harder than
   (b) harder than
   (c) not harder than
   (d) softer than
   (e) as soft or softer than

8. A relation, R, among kinds of minerals is written 'harder than'. The complement in this set, $\bar{R}$, is ______________________
   (a) as hard or harder than
   (b) harder than
   (c) softer than
   (d) not harder than
   (e) not as hard as

ANSWERS

1. b	3. f	5. c	7. e
2. e	4. a	6. b	8. d

End of Part III

Index

About the Author

Morton L. Schagrin received his B.A., B.S., and M.A. degrees from the University of Chicago and his Ph.D. from the University of California (Berkeley). Currently professor of philosophy and chairman of the philosophy department at the State University of New York (Fredonia), he has also taught at Denison University and the University of Florida (Gainesville), and in 1966 he spent a year as a research associate at Harvard University. Professor Schagrin is a member of the Philosophy of Science Association, the History of Science Society, and the American Philosophical Association. His articles and reviews have appeared in numerous professional journals such as *Philosophy of Science*, the *British Journal for the Philosophy of Science*, and the *American Journal of Physics*. At present he is doing research on the experimental evidence for the pressure of light.